Challenging Cases in Diagnostic Clinical Microbiology

James J. Dunn • Marc Roger Couturier
Audrey N. Schuetz
Editors

Challenging Cases in Diagnostic Clinical Microbiology

Advanced Strategies and Techniques

Springer

Editors
James J. Dunn
Department of Pathology
Texas Children's Hospital
Houston, TX, USA

Marc Roger Couturier
Department of Pathology
University of Utah/ARUP Laboratories
Salt Lake City, UT, USA

Audrey N. Schuetz
Department of Laboratory Medicine and Pathology
Mayo Clinic
Rochester, MN, USA

ISBN 978-3-031-65091-8 ISBN 978-3-031-65092-5 (eBook)
https://doi.org/10.1007/978-3-031-65092-5

This Springer imprint is published by the registered company Springer Nature Switzerland AG
The registered company address is: Gewerbestrasse 11, 6330 Cham, Switzerland

We are extremely grateful to all of the teachers, mentors, and role models we've had over the years and the time and spirit they afforded to inform and educate us.

Preface

The methods for diagnosis of infectious diseases continue to advance technologically at a rapid pace, in most instances to provide actionable information in a timely manner to influence patient management. For many syndromes clinicians might encounter, new testing approaches and information accrue rapidly and it is important for all involved in patient care to stay abreast of the latest advances in diagnostics and the ever expanding etiologies of different types of infections. With these advancements, it becomes necessary for healthcare providers to have a firm grasp of the intricacies and nuances of the performance and utility of these newer test methods and how the results should be interpreted for various patient populations. Technology is not always the panacea we hope it will be. These all come with their own unique limitations and advantages, and this text is intended to communicate and educate on these subtleties to those who are using or intend to use such modalities in their care of patients. We aim to disseminate as broadly as possible to a wide audience the information and real-world experiences of how best to apply laboratory diagnostic tests for infectious diseases in order to facilitate the diagnosis and treatment of patients in the most appropriate or ideal manner.

Clinical microbiologists, pathologists, infectious disease and community healthcare providers, medical residents, fellows, and other trainees will find this text useful for facts and information related to the practice of clinical microbiology and how laboratory testing can be used to provide data impactful to diagnose, treat, and manage patients with infectious diseases—information that is important for individual patient outcomes as well as public health in general. The overview sections provide specifics on the methodologies considered current standard of care in the various areas of laboratory testing, including appropriate specimen selection and transport requirements, immunodiagnostics, and methods for culture-based detection, direct staining, molecular-based detection, and molecular and culture-based antimicrobial susceptibility testing. The individual cases impart application of these laboratory skills and, we hope, allow for healthcare professionals to more easily understand

and navigate the challenging area of infectious disease diagnostics in order to assure successful outcomes for their patients.

Houston, TX, USA — James J. Dunn
Salt Lake City, UT, USA — Marc Roger Couturier
Rochester, MN, USA — Audrey N. Schuetz

Acknowledgments

We would like to thank the technical staff in our laboratories for their tireless effort and dedication to high-quality patient care and for their expertise in pursuit of providing timely, accurate information to diagnose and treat infectious diseases.

Contents

Contributors

Megan H. Amerson-Brown, PhD, D(ABMM), CIC, MLS(ASCP) Department of Pathology, UAB Heersink School of Medicine, The University of Alabama at Birmingham, Birmingham, AL, USA

Cameron A. Brown, PhD, D(ABMM) Departments of Pathology & Immunology and Pediatrics, Baylor College of Medicine, Houston, TX, USA

Marc Roger Couturier, PhD, D(ABMM) Department of Pathology, University of Utah/ARUP Laboratories, Salt Lake City, UT, USA

James J. Dunn, PhD, D(ABMM) Department of Pathology, Texas Children's Hospital, Houston, TX, USA

Department of Pathology & Immunology, Baylor College of Medicine, Houston, TX, USA

Santiago Delgado Fernandez, MD Pathology Consultant, Caracas, Venezuela

Angela Ma, PhD Public Health Ontario, Department of Laboratory Medicine and Pathobiology, University of Toronto, Toronto, ON, Canada

Anisha Misra, PhD, D(ABMM) Department of Pathology, Cleveland Clinic/ Lerner College of Medicine, Cleveland, OH, USA

Audrey N. Schuetz, MD, MPH, D(ABMM) Department of Laboratory Medicine and Pathology, Mayo Clinic, Rochester, MN, USA

Chapter 1
Bacteriology and Mycobacteriology Cases

Audrey N. Schuetz, Marc Roger Couturier, Angela Ma, Santiago Delgado Fernandez, Cameron A. Brown, Megan H. Amerson-Brown, and James J. Dunn

A. N. Schuetz (✉)
Department of Laboratory Medicine and Pathology, Mayo Clinic, Rochester, MN, USA
e-mail: Schuetz.Audrey@mayo.edu

M. R. Couturier
Department of Pathology, University of Utah/ARUP Laboratories, Salt Lake City, UT, USA
e-mail: Marc.Couturier@aruplab.com

A. Ma
Public Health Ontario, Department of Laboratory Medicine and Pathobiology, University of Toronto, Toronto, ON, Canada
e-mail: Angela.Ma@oahpp.ca

S. D. Fernandez
Caracas, Venezuela

C. A. Brown
Departments of Pathology & Immunology and Pediatrics, Baylor College of Medicine, Houston, TX, USA
e-mail: Cameron.Brown@bcm.edu

M. H. Amerson-Brown
Department of Pathology, UAB Heersink School of Medicine, The University of Alabama at Birmingham, Birmingham, AL, USA
e-mail: mamersonbrown@uabmc.edu

J. J. Dunn
Department of Pathology, Texas Children's Hospital, Houston, TX, USA

Department of Pathology & Immunology, Baylor College of Medicine, Houston, TX, USA
e-mail: jjdunn@texaschildrens.org

J. J. Dunn et al., *Challenging Cases in Diagnostic Clinical Microbiology*,
https://doi.org/10.1007/978-3-031-65092-5_1

Overview of Diagnostic Bacteriology and Mycobacteriology Methods

Diagnostic Bacteriology Methods

The detection of bacteria in clinical specimens is a process that has been performed for over a hundred years. The advances in technology have developed fast and furious in the past 10–15 years, compared to the previous 100 years. Most classical testing was focused on direct staining of bacteria from clinical specimens and subsequent growth of the offending bacteria on a solid media. Identification was attained with subsequent testing for arrays of biochemical reactions and metabolic processes. These processes could take days, sometimes over a week for a definitive identification to be determined (and those identifications may not have been entirely accurate). More recent technologies using molecular detection (including DNA, protein, carbohydrates, etc.) have allowed for rapid identification with or without adjunct biochemical reactivity, and some methods obviate the need for bacterial culture all together (so-called culture-independent diagnostics tests [CIDT]).

Gram Stain

The Gram stain methodology is the routine diagnostic bacteriology stain for microscopic visualization of bacteria in primary specimens and cultured media. Developed by Danish microbiologist Hans Christian Gram in 1884, the principle of the Gram stain is based on the differences in peptidoglycan cell wall thickness and components among bacteria, allowing their broad classification into two groups. Gram-positive organisms possess thick peptidoglycan cell walls that are abundant in teichoic acids, which are negatively charged polymers comprised of glycerol phosphate or ribitol phosphate molecules linked to carbohydrates through phosphodiester bonds. Conversely, Gram-negative organisms possess relatively thinner peptidoglycan cell walls and an outer lipopolysaccharide membrane.

The Gram stain procedure consists of five main steps, beginning with affixing bacteria to a glass slide either from primary specimens or culture media. In instances of Gram staining bacterial colonies from solid media, they should be emulsified using sterile saline or water into a thin solution first. After drying, slide material should then be fixed using 70–95% alcohol or gentle heat to prevent material from being washed off during staining. Slides are flooded with the basic dye crystal violet (hexamethyl pararosaniline chloride) for 30 s before briefly rinsing with water. A dilute iodine solution is then applied to the slide for 30 s, which forms a complex with crystal violet by replacing the chloride present in the molecular structure of crystal violet with iodine. Addition of Gram's iodine serves as a mordant, facilitating penetration of the crystal violet-iodine complex into the peptidoglycan cell wall of bacteria. At this point, all material present on the slide is a deep purple to dark blue color. The next step of the Gram stain is critical for differentiation of Gram-positive and Gram-negative organisms. Slides are decolorized with acetone or absolute alcohol and rinsed with water; at this point Gram-positive organisms will retain the crystal violet-iodine complex due to the increased thickness of their peptidoglycan cell wall. Gram-negative organisms are unable to retain the dye due to the damage sustained to their thinner peptidoglycan cell walls from the acid-alcohol solution, which results in the dye complex washing out. Following decolorization, Gram-negative organisms and any other unstained cellular elements are counterstained with safranin or dilute carbolfuchsin dye for 30 s to appear pink to red in color. Slides should be gently rinsed after counterstaining and dried before evaluating under a bright-field microscope.

In addition to the Gram reaction, the shape of individual cells and arrangement of bacteria aid in providing diagnostic information and directing further testing. Bacteria can be broadly categorized into groups based on their shapes and arrangements, including cocci, bacilli, coccobacilli, and diplococci (Fig. 1.1). Detailed

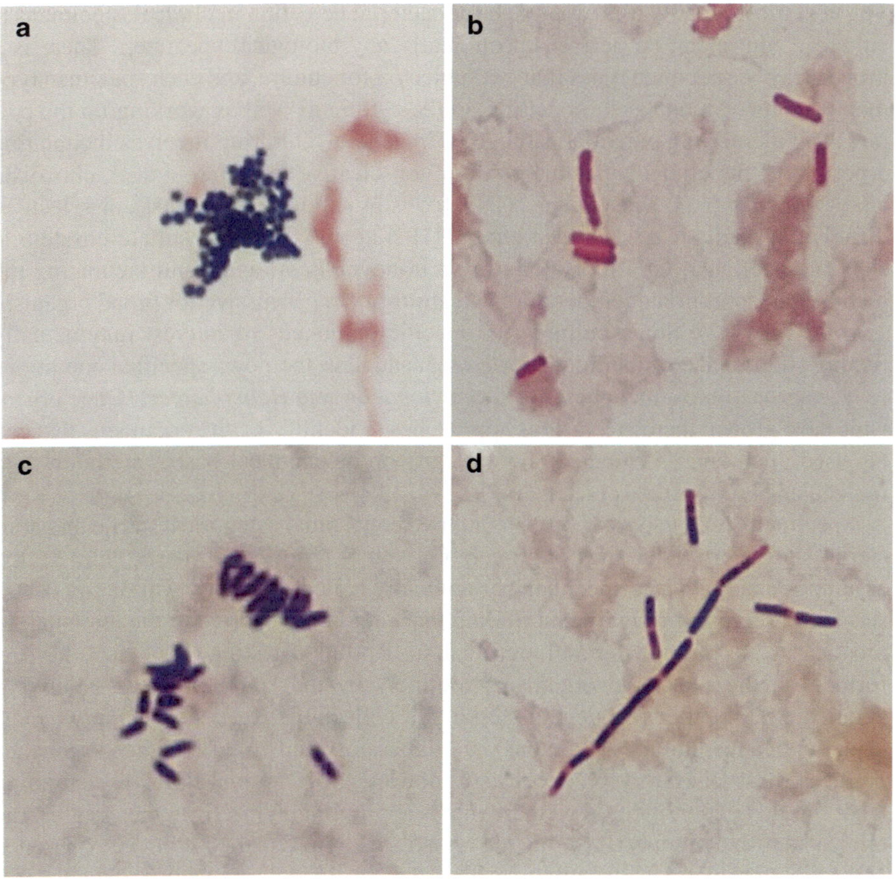

Fig. 1.1 Direct Gram stains from positive blood cultures of (**a**) *Staphylococcus aureus*, Gram-positive cocci in clusters; (**b**) *Escherichia coli*, Gram-negative bacilli; (**c**) *Corynebacterium* spp., Gram-positive bacilli in palisades; (**d**) *Paenibacillus* spp., Gram-variable spore-forming bacilli

descriptions of Gram stains are incredibly useful for guiding identification of bacteria, such as Gram-positive cocci in clusters correlating with *Staphylococcus* species and Gram-negative spiral-shaped bacilli being suggestive of *Helicobacter* species. Important considerations of the Gram stain are to avoid potential over- or under-decolorization that can result in falsely reporting Gram-positive bacteria as Gram-negative and vice versa. Bacteria also have varying avidities to the dyes used in the Gram stain, which can result in Gram-variable or faintly staining organisms despite proper staining technique. In situations of prior antimicrobial exposure or abundant dead organisms, Gram stains may appear atypical. Hence, it is important to always correlate the Gram reaction with the morphology of the bacteria, colony morphology on solid media if available, specimen type, and clinical scenario.

Agar-Based Culture

The next most broadly used approach for bacterial detection in clinical specimens is culture. Culture can be performed on nearly any biological specimen. There is a broad array of specimen types that are subjected for culture, and each specimen type may have its own nuances for setting up the culture as well as working up the cultures and identifying potential pathogens. In general, culturing involves inoculating a portion of the clinical specimen onto a nonselective (e.g., blood agar, chocolate agar), selective (e.g., MacConkey, CNA [colistin nalidixic acid agar]), or selective/differential media (e.g., Hektoen enteric [HE] agar, TCBS [thiosulfate-citrate-bile salts-sucrose] agar, CIN [cefsulodin-irgasan-novobiocin] agar) and incubating the media at an appropriate temperature and atmosphere permissive for broad organism growth (Fig. 1.2). Some cultures for specific pathogens of interest may actually require additional considerations for growth and have their own specified conditions (e.g., microaerobic atmosphere for *Campylobacter* and *Helicobacter*). Once organisms have grown, there are various approaches to identifying the organisms that can be used including serological typing, growth/biochemical-based methods, and molecular-based methods (see below).

Specimens can be loosely broken into two conceptual categories: sterile and non-sterile sites. For sterile sites (e.g., cerebral spinal fluid [CSF], tissue, blood), the assumption is that if the specimen is aseptically collected, the growth of any bacteria should represent a significant finding that should be reported to the clinician. In most cases, accompanying antimicrobial susceptibility testing (AST) will be performed on these isolated organisms routinely by the laboratory (see below for detailed AST methods). For non-sterile site collections (e.g., stool, skin/wounds, urine, upper respiratory specimens), the laboratory must make clinical decisions based on best laboratory practices and knowledge of "normal flora" to determine what should be identified, as well as what is appropriate for AST. In many cases, the clinical staff may request AST testing on isolates based on their clinical experience and patient evaluation. Consultations between the clinical care team and the laboratory should occur whenever possible.

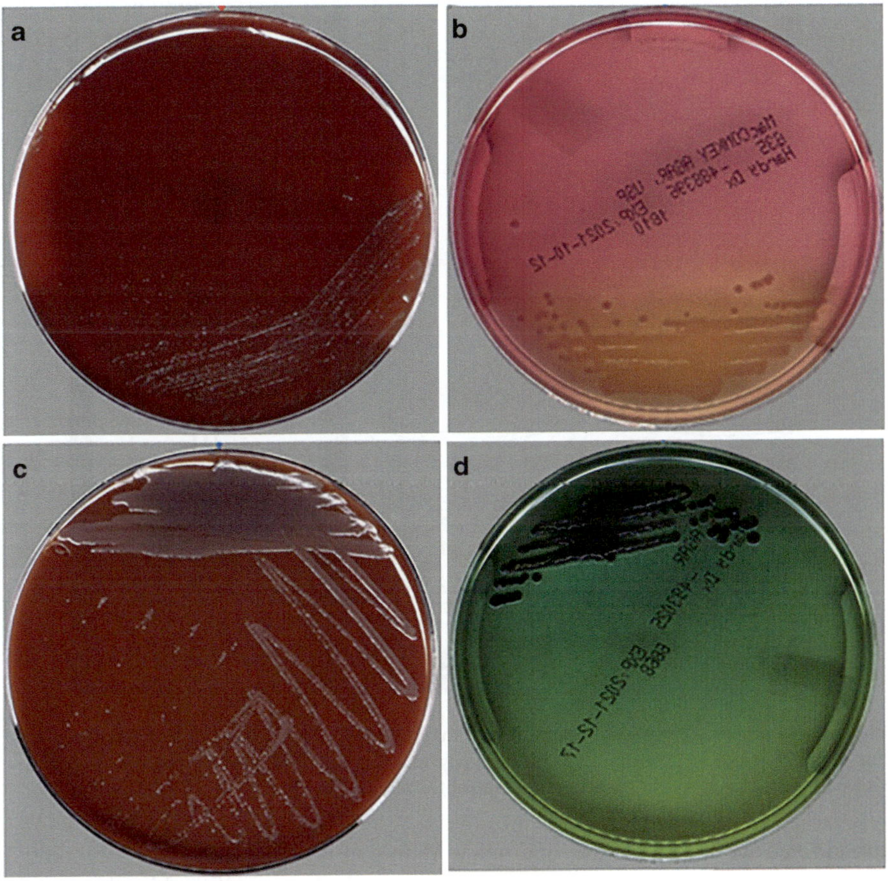

Fig. 1.2 Common culture media used in clinical microbiology laboratories. (**a**) Blood agar plate with mixed Gram-positive flora. (**b**) MacConkey plate with non-lactose fermenting Gram-negative rods. (**c**) Chocolate agar with coagulase-negative *Staphylococcus*. (**d**) Hektoen enteric agar with *Salmonella* spp.

Blood Culture

Traditionally, blood was incubated in tubes and manually examined every day by Gram staining an aliquot of the blood to look for the presence of bacteria. This was both time- and labor-consuming as well as a risk for contamination. Conventional blood cultures consist of an aerobic and anaerobic blood bottle for each unique blood collection, and multiple collections should be obtained from different anatomical sites within a short period (e.g., left arm and right arm). Contemporary blood cultures are performed using special blood culture bottles which are monitored continuously for evidence of growth using biochemical indicators. Continuous blood culture monitoring systems are available from multiple manufacturers and allow more rapid and timely detection of bacterial growth as well as a more

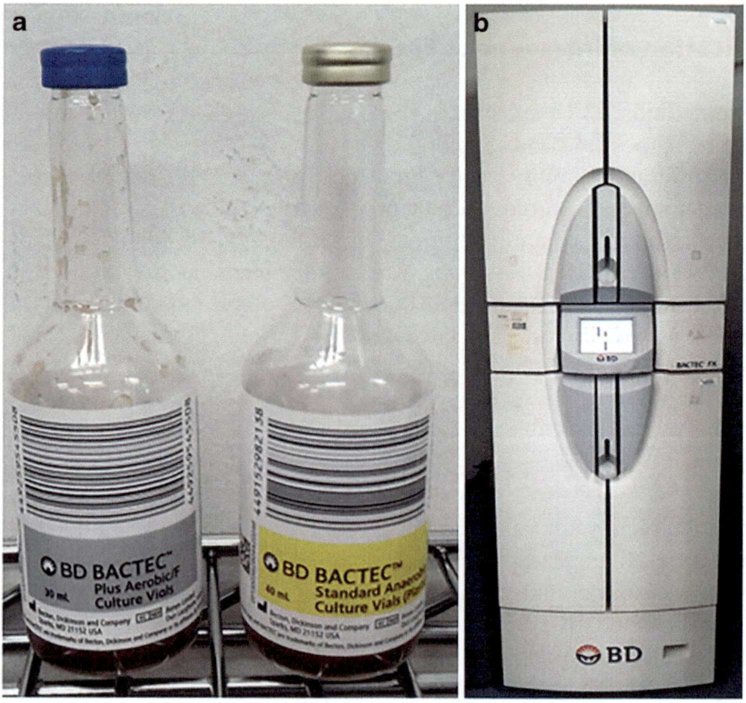

Fig. 1.3 Blood culture bottles for aerobic and anaerobic blood culture (**a**). The two bottles constitute a single collection. Automated continuously monitoring blood bottle incubator (**b**). Internal racks are maintained within the doors to control for monitoring and appropriate temperature

streamlined workflow for the laboratory. With continuously monitoring blood culture systems, a biochemical reporter is coated on the bottom of the blood bottle and the instrument detects the presence of bacterial growth through a chemical reaction with some by-products of the bacterial metabolism (e.g., internal fluorescent CO_2 sensors detect the metabolic activity of bacteria growing in a bottle) (Fig. 1.3a). The instrument scans the bottles continuously for the presence of a "signal" which is produced when a threshold of bacterial growth/metabolism is reached (Fig. 1.3b). The technologist then removes the bottle from the instrument immediately, removes an aliquot using a sterile syringe, performs a Gram stain (see above), and plates the blood culture aliquot to appropriate media, depending on the Gram stain (e.g., a Gram-negative organisms will be plated to nonselective media and MacConkey whereas Gram-positive organisms will be plated to nonselective media and CNA).

The blood culture bottle is not typically returned to the instrument as it can only "flag" one time. Laboratories may choose to hold positive bottles at room temperature or return them to a traditional 35–37 °C incubator for further workup as needed later. Some instances may justify a subsequent workup after additional incubation, including but not limited to: (1) the Gram stain reveals two organisms but only one grows on the initial subculture; (2) one organism grows on initial subculture but a

CIDT method indicates another organism may be present; (3) the patient is in a risk group for polymicrobial bacteremia with fastidious organisms (e.g., pan-mucositis, abdominal perforation) and only one (typically rapid growing organism) was seen in the Gram stain or culture; (4) the physician has requested an extended incubation for suspicion of a fastidious organism but another organism has already grown.

Any growth seen on initial subculture or subsequent subcultures should be identified by appropriate and available methods in the laboratory and AST should be performed if appropriate. Certain organisms are typically associated as skin contaminants (*Micrococcus,* coagulase-negative *Staphylococcus,* diphtheroids) and should be worked up only if there are extenuating clinical situations. In general, the presence of these aforementioned organisms in a single blood draw indicates contamination with skin flora. The presence of one of these organisms in multiple blood draws could represent a true infection, or consecutive poor collections. Clinical consultation is appropriate in these scenarios.

Biochemical Identification

Prior to the adoption of newer technologies such as matrix-assisted laser desorption ionization-time-of-flight mass spectrometry (MALDI-TOF MS) and 16S rRNA gene sequencing, definitive identification of bacteria was primarily achieved by phenotypic biochemical testing. However, identification of bacteria to a species level solely by biochemical tests is highly laborious and significantly increases turnaround time for cultures compared to modern methods. Many of the biochemical tests assess whether an organism can utilize different carbohydrates through fermentative or oxidative pathways, enzymatic activity, or presence or absence of metabolic end products. In brief, biochemical tests can be classified into two broad categories, rapid "spot" biochemical tests and those that require incubation, typically for 18–24 h. Due to the advantage of rapid results from spot biochemical tests, they are still routinely used in the laboratory and can provide complete genus and species identification for some organisms such as *Staphylococcus aureus*.

Two spot biochemical tests that are mainstays in diagnostic bacteriology are the catalase and oxidase tests. The catalase test determines if an organism produces the enzyme catalase, which decomposes hydrogen peroxide to oxygen and water. Isolated colonies from solid media are picked and submerged in a drop of hydrogen peroxide. Enzymatic activity of catalase results in the rapid formation of bubbles that can be visualized macroscopically (Fig. 1.4). Catalase production allows for differentiation of the Gram-positive *Micrococcaceae* family that are catalase positive from the catalase negative *Streptococcaceae* family. The oxidase test is a rapid biochemical test that evaluates whether a bacterium's cytochrome oxidase enzymes are capable of oxidizing an artificial electron acceptor in the presence of atmospheric oxygen. Isolated colonies are spotted onto filter paper saturated with the phenylenediamine reagent tetra-methyl- p-phenylenediamine dihydrochloride. Bacteria with cytochrome oxidase activity will cause a color change in the phenylenediamine reagent from colorless to a dark blue or purple due to the production of

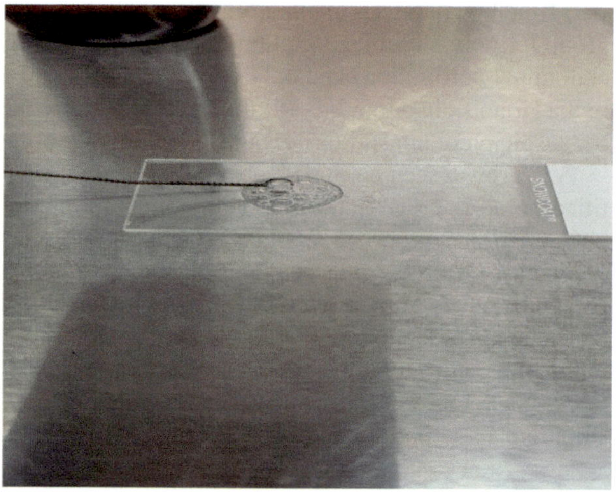

Fig. 1.4 Positive catalase activity in *Staphylococcus aureus* reacts with hydrogen peroxide to cause the production of oxygen bubbles

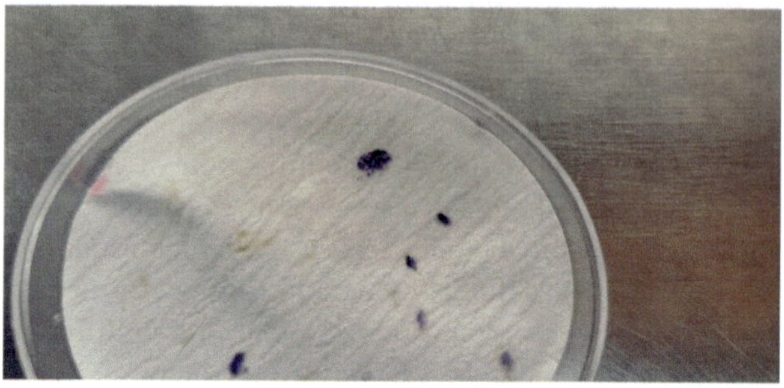

Fig. 1.5 Positive cytochrome oxidase reaction by *Pseudomonas aeruginosa* results in the formation of a purple compound when a colony is rubbed onto filter paper saturated with phenylenediamine reagent

indophenol blue (Fig. 1.5). Among Gram-negative bacteria, all *Enterobacterales* with the exception of *Plesiomonas shigelloides* are oxidase negative. Oxidase-positive non-*Enterobacterales* Gram-negative bacilli commonly encountered include *Campylobacter* spp., *Aeromonas* spp., *Vibrio* spp., and *Pseudomonas aeruginosa*.

Many biochemical properties of bacteria cannot be rapidly assessed like catalase and oxidase activity and typically require incubation of bacteria with various substrates, compounds, and pH indicators in agar or broth for phenotypic reactions to be detectable. Historically, biochemical tests were performed by inoculating bacteria into individual test tubes. While this method is still performed in laboratories, biochemical tests are also now commercially available in panels that contain multiple small reaction wells or tubes for different biochemical characteristics that allow targeted identification of bacteria. Furthermore, semiautomated and automated platforms for biochemical identification of bacteria are now widely used in diagnostic bacteriology laboratories, reducing the manual labor of inoculating and reading biochemical tests.

16S rRNA Sequencing

Gene sequencing has become a standby method for bacterial identification in the clinical microbiology laboratory. Unlike growth-based methods, sequencing is more objective and does not require optimal growth, traditional morphological features to be seen, or even living organisms. The most commonly targeted gene for bacterial identification is 16S rRNA (or 16S rDNA). The 16S rRNA gene is approximately 1500 base pairs in length and codes for part of the 30S ribosome. It is a highly conserved housekeeping gene in prokaryotes, but shows taxonomic diversity among genera and species. Traditionally, only the 5′ 500–base pair region of the 16S rRNA gene has been targeted by Sanger sequencing, which allows for robust specification of most organisms. Some exceptions exist however, and this region is not a universal solution for all species-level identifications. The gene can also be multicopy, which can lead to difficulty determining a precise species when base pair substitutions exist in only one copy of the gene. For some organisms such as *Mycobacterium*, 16S rRNA is not discriminatory and the *rpo*B gene is a better target for sequencing.

Once sequence data has been obtained from a bacterial isolate or direct clinical specimen, the laboratory will query a database of known bacterial gene/genome sequences against the sequence generated from the unknown organism. These databases can be privately curated or publicly available. There are strengths and weaknesses to both database types. Large public databases are free to use and contain a broader diversity of sequences, but there is limited curation and thus the quality of the database cannot be predicted. With privately curated databases, there is a fee-for-service involved and less sequence and organism diversity, but a better assurance of database integrity. Nucleotide sequences are reported as a percent identity. The percent identity refers to the number of identical nucleotides shared by the unknown organism and the reference sequences, divided by the total number of nucleotides sequenced. Genera can be reliably identified at greater than or equal to 97% identity. Species-level identification is generally reliable at greater than 99% identity. In some cases, two species may be identified with the same percent identity score, necessitating differentiating biochemicals or other traditional microbiological features. The so-called slash-calls may be required for some species that cannot be separated by any conventional biochemicals or techniques.

Matrix-Assisted Laser Desorption Ionization-Time-of-Flight Mass Spectrometry (MALDI-TOF MS)

Mass spectrometry has long since been utilized in diagnostic chemistry and toxicology laboratories for the detection and quantification of small molecules. Although relatively recent compared to other divisions of laboratory medicine, mass spectrometry, specifically MALDI-TOF MS, has now become widely adopted for rapid identification of bacteria in diagnostic microbiology. Specifically, there are two FDA-cleared MALDI-TOF MS systems for use in the United States that diagnostic bacteriology laboratories employ for identification of aerobic and anaerobic bacteria. Other than the upfront costs of purchasing the instrumentation, MALDI-TOF MS is an inexpensive method to rapidly and accurately identify bacteria.

The MALDI-TOF MS system enables identification of bacteria to a genus and/or species level through the analysis of bacterial cell's protein composition. Bacteria grown on solid agar can be prepared for MALDI-TOF MS identification by two methods, the direct spot or with extraction. In the direct spot method, bacterial colonies are smeared onto the MALDI-TOF MS target plate and coated with a solution of an energy-absorbent matrix that entraps and co-crystallizes the sample when dried on the plate. The target plate is then loaded onto the instrument and a laser fires a beam of energy onto each spot to ionize the sample into positively charged ions. Individual ions are then accelerated and separated from each other on the basis of their mass-to-charge ratio (*m/z*), which is detected and measured using the TOF MS analyzer to generate a spectral fingerprint. The spectral fingerprint of the ionized peptides from the isolate is compared to a reference database containing multiple spectra of different bacteria. Based on how closely the spectral fingerprint of the isolate is to an entry in the reference database, a bacterial identification and corresponding quality score is determined. In situations where the bacterial cell structure is more complex or when the direct spot method fails to produce a reliable bacterial identification, extraction may be necessary. Through the use of chemical- and mechanical-assisted cell lysis, proteins are released from the cell into solution, improving the resolution of the protein spectrum generated.

Antimicrobial Susceptibility Testing

Phenotypic antimicrobial susceptibility testing (AST) is a central component of the diagnostic bacteriology laboratory and plays a significant role in appropriate therapy interventions and patient management in cases of bacterial infections. Furthermore, the data generated from a laboratory's AST is incredibly useful in guiding antimicrobial stewardship for an institution. There are numerous factors to consider when determining whether AST is appropriate for an organism(s) and what the most suitable testing method should be. For bacteria that have predictable susceptibility patterns, such as *Streptococcus pyogenes* and beta-lactam agents, AST is typically not encouraged. Laboratorians should be familiar with appropriate sample collection procedures, evaluating predominant organisms of clinical significance from normal microflora, identifying contraindicated AST requests that may result in adverse patient outcomes, and recognizing clinical scenarios associated with immunocompromised hosts that may warrant susceptibility testing for bacteria that may not be significant in immunocompetent hosts. From a technical standpoint, an important consideration for AST is to have a proficient understanding of the methods and guidelines for testing and interpretations, respectively, as outlined by the FDA and Clinical and Laboratory Standards Institute (CLSI), the latter being an expert-driven organization that develops standards for laboratory testing. In routine diagnostic bacteriology, phenotypic AST predominantly focuses on determining the lowest concentration in µg/mL of an antimicrobial agent required to kill 99.9% of a bacterial isolate in vitro, which is known as the minimum inhibitory concentration (MIC). From the MIC values obtained, breakpoints correlating with susceptible, intermediate, and resistant results are available for many bacteria that help guide therapy decisions. There are various methods for phenotypic AST testing that laboratories may perform in-house or send to larger reference laboratories if necessary.

Standardization of the bacterial inoculum is critical regardless of the AST method performed. For aerobic and anaerobic bacteria, a 0.5 McFarland standard that corresponds to approximately 1.5×10^8 colony forming units/mL is required. Bacteria should firstly be grown in appropriate incubation conditions according to CLSI guidelines to achieve log-phase growth, and then suspended in a tube of saline or sterile water. Historically, laboratories established a 0.5 McFarland standard by visually inspecting the optical density of the suspension against a striped card and comparing the turbidity to a calibrated control tube. Modern electronic densitometers have now largely replaced the necessity of manually inspecting tubes. Regardless of the AST method performed, appropriate quality control processes such as control strains for the antimicrobial agents tested and purity check plates should always be included to monitor test performance.

Broth Microdilution (BMD) BMD is a semiquantitative, modern approach to broth macrodilution tests, which historically used test tubes. Doubling dilutions of antimicrobial agents are frozen or lyophilized in the wells of a sterile microtiter plate, that are then resuspended upon addition of Mueller-Hinton broth and a 0.5 McFarland standard of bacterial inoculum. A variety of panels are FDA-cleared and commercially available for groups of organisms based on Gram stain reaction, growth requirements, and incubation conditions. Standardized methods and guidelines for BMD of different organisms are published by the CLSI and should be followed for proper interpretation of MIC values obtained. Bacteria in wells without antimicrobial agent, or when they can overcome the effects of the antimicrobial concentration, are visible as a button of growth >2 mm in size at the bottom of the well. For BMD, MIC values are read at the lowest concentration of a given antimicrobial agent where no button of bacterial growth is observed (Fig. 1.6). Caution should be taken when reading MIC values for BMD as microtiter plates contain multiple antimicrobial agents, all with different ranges of doubling dilutions. MIC values are accurate for a bacterial isolate ±1 doubling dilution, and it is important for laboratorians and physicians to not attribute this natural variation in MIC as changes in susceptibility. Furthermore, trailing endpoints may occur, particularly with bacteriostatic agents, and buttons <2 mm in size should be disregarded and not falsely interpreted as the MIC value.

BMD is considered the reference AST method for the majority of aerobic bacteria but among anaerobic bacteria, guidelines are established only for the *Bacteroides fragilis* group due to poor correlation of MIC values for other anaerobic bacteria with agar dilution. Manual visual inspection for the presence or absence of a button by a trained laboratorian is the reference method for reading BMD microtiter plates, but there are various automated instruments that can reduce the manual labor

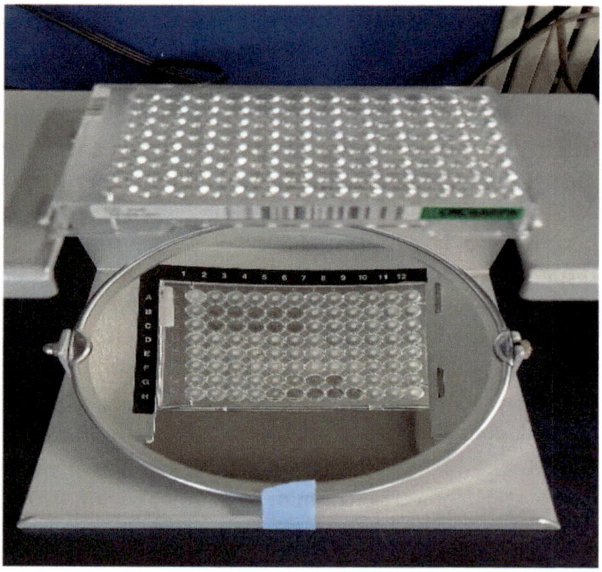

Fig. 1.6 Broth microdilution microplate on a mirrored viewbox for manual reading of the minimum inhibitory concentration following inoculation of a 0.5 McFarland standard of bacterium and incubation of the plate

involved in BMD. Inoculation, incubation, and plate reading instruments can be implemented in laboratories to improve BMD efficiency and workflow, and there are a growing number of FDA-cleared combined bacterial identification and phenotypic AST systems based on BMD that many diagnostic bacteriology laboratories now use.

Disk Diffusion (Kirby-Bauer) The disk diffusion test is based on inoculating a defined volume of a 0.5 McFarland standard of bacterium onto a plate of Mueller-Hinton agar ± additional supplements (as directed by the CLSI guidelines). The inoculum is spread in three different directions across the entire surface of the agar plate to enable confluent growth of the bacterium when incubated. Following inoculation of the bacteria, sterile filter paper disks impregnated with antimicrobials are placed onto the plate. After incubation, the zone diameter of bacterial growth inhibited by the antimicrobial is measured in millimeters (Fig. 1.7). This zone of inhibition is semiquantitative as the diameter does not correlate with an MIC specifically, but there are breakpoints for susceptible, intermediate, and resistant interpretations available for many bacteria.

Zones of growth inhibition can be influenced by various factors, hence standardization in the testing method is important. Agar depth, pH, and composition may detrimentally affect the zone diameter, potentially causing bacterial susceptibility to be falsely interpreted. In addition, failure to follow incubation conditions and

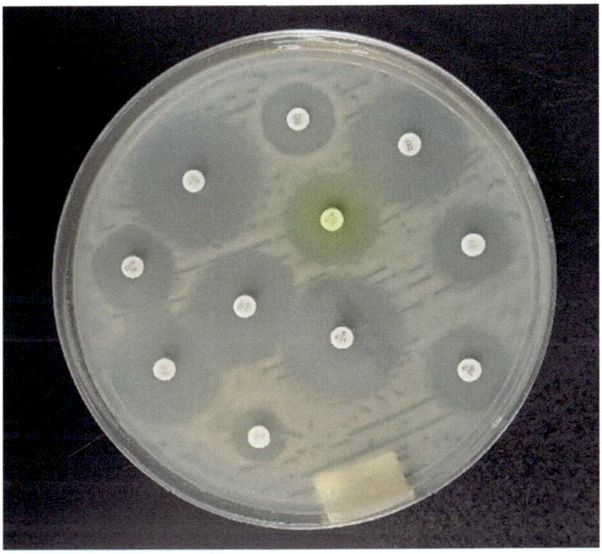

Fig. 1.7 Disk diffusion test with multiple filter paper disks impregnated with antimicrobial agents placed on a Mueller-Hinton agar plate inoculated with a 0.5 McFarland standard of bacterium

improper number or composition of disks placed onto plates can cause inaccurate disk diffusion results. In situations of mixed bacterial populations, the zones of inhibition may resemble hetero- or homoresistant subpopulations, which appear as a haze of growth approaching the disk or isolated colonies within a defined zone of inhibition, respectively. Disk diffusion testing offers the advantages of multiple antimicrobial agent-disk combinations on a single agar plate, ease of set up, and that it's relatively inexpensive. However, limitations of the method include the lack of a true MIC value and the inability to fully automate the method although there are now instruments available that digitally measure the zone diameters on computer systems.

Gradient Diffusion Gradient diffusion/Epsilometer tests are performed similarly to disk diffusion but instead of using paper disks with a fixed concentration of antimicrobial agents, inert, non-porous plastic or paper carrier strips are saturated with predefined continuous antimicrobial gradients. The underside of the strip contains the antimicrobial gradient, whereas the surface is printed with an MIC interpretive scale. Gradient diffusion strips are placed onto Mueller-Hinton agar plates that have been inoculated with 0.5 McFarland standards of bacteria and incubated per standardized guidelines. Upon placing the strip onto the plate, the antimicrobial agent immediately diffuses into the agar, creating a continuous and exponential gradient of antimicrobial concentration along the strip. After incubation, the MIC is deter-

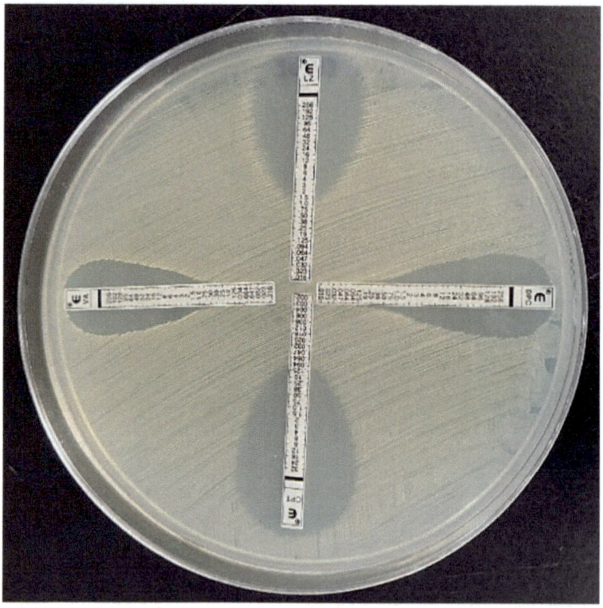

Fig. 1.8 Gradient diffusion test strips of linezolid (top), daptomycin (right), ceftaroline (bottom), and vancomycin (left) on a Mueller-Hinton agar plate inoculated with a 0.5 McFarland standard of *Staphylococcus aureus*

mined at the elliptical intersection of the zone of inhibition edge and strip (Fig. 1.8). Typically, bactericidal agents will produce a sharp endpoint on gradient strips due to the complete inhibition of growth of the organism. On the other hand, trailing endpoints with hazy bacterial growth beyond the endpoint may be observed with gradient diffusion strips impregnated with bacteriostatic agents. MIC values should thus be read at 80–90% inhibition for bacteriostatic agents. Gradient diffusion strips are FDA-cleared for use with common bacteria and antimicrobial agents and recommended for AST of fastidious organisms, but are not considered a reference test method.

Agar Dilution For most anaerobic bacteria and a selection of fastidious organisms such as *Helicobacter pylori*, agar dilution is the gold standard method for AST. Agar dilution is a quantitative method that involves inoculating 0.5 McFarland standards of a maximum of 30 different bacterial isolates onto freshly prepared agar plates with predefined concentrations of antimicrobial agents. Antimicrobial solutions are added to molten agar prior to pouring into petri dishes to solidify. Typically, at least six doubling dilutions per antimicrobial agent are required, with two dilutions being below the susceptible breakpoint. Following incubation, plates are examined for bacterial growth. The plate with the minimum concentration of antimicrobial agent that completely inhibits growth of the isolate is considered the MIC; single colonies or haze due to dried inoculum should be ignored. Among phenotypic AST methods, agar dilution is the most labor intensive and time consuming; hence, it is only routinely performed by a select number of reference laboratories. Use of a steer's replicator that allows multiple bacterial isolates to be inoculated onto the agar plates at once can improve efficiency, but there are no automated instruments available for agar dilution.

Nucleic Acid Detection and Syndromic Panels

Multiplex molecular panels have become widely available for clinical detection of pathogens from various specimen sources. These panels can vary in terms of the number of organisms detectable, the throughput of the testing instrument, the turnaround time, and the complexity of the testing. The technologies are generally based on real-time PCR; however, some assays may incorporate nested PCR, signal amplification steps, microarray read-outs, or hybridization to specific fluorescent probes with signal amplification but not target amplification. Some of these products can be used one specimen at a time, near point of care, with approximately an hour turnaround time. Other products require batching of dozens to hundreds of specimens and upwards of 6–8 h of laboratory testing time to generate results. In general, these tests are considered highly sensitive and specific, although nuanced and specific limitations exist for each product. Users should consult the manufacturer's product inserts for specifications of the assay and each analyte as well as the specific intended use statement and any requirements for adjunct testing. Primary research literature should also be consulted to evaluate "real world" performance studies to further understand the advantages and limitations of these products.

Blood Blood culture pathogen identification panels are commercially available for use from direct blood specimens as well as post-blood culture bottle incubation. For products requiring preincubation and flagging by a blood culture instrument, a Gram stain should be performed before performing the multiplex panel as part of the intended use requirements. Some products are broken into specific panels for Gram-negative organisms, Gram-positive organisms, and yeasts. The Gram stain will dictate which panel or panels should be used on that specific blood culture bottle. Other products test for all the targets (Gram-positive, Gram-negative, and yeast) at the same time regardless of the Gram stain result, but a Gram stain is still required. These products generally target the most common causes of bacteremia/fungemia and sepsis and may include several genes that predict antimicrobial resistance. A critical point to consider is that the tests are not a substitute for culture and conventional AST and instead are intended to provide early and actionable results to treating physicians. Several studies have shown that these tests only have clinical value if the results are verbally relayed to a physician or antimicrobial stewardship pharmacist in real time, and there is subsequent review and refinement of empiric antimicrobials where appropriate.

Stool The gastrointestinal pathogen panels that are commercially available are much more diverse than is seen with blood culture identification panels. Specimens are generally stool in enteric transport media or non-formalin-containing fixatives. Unlike the blood culture identification panels, these panels have been aggressively marketed as replacements for conventional detection methods such as bacterial stool culture and viral/protozoan antigen detection and augment the need for stool ova and parasite examinations. Some products only target common gastrointestinal bacterial pathogens. Other products target common viruses and common protozoa as standalone minipanels which serve as a complement to the standalone bacterial panel. A third approach embraced by several manufacturers involves a more syndromic detection where gastrointestinal pathogenic viruses, bacteria, and common protozoa are detected simultaneously in a single test. Importantly, these assays do not cover all possible causes of infectious diarrhea. In patients who test negative on a syndromic panel, additional conventional testing (e.g., ova and parasite examination, special stool stains for coccidia and microsporidia) must be considered based on the patient's clinical signs and symptoms.

CSF Multiplex molecular panels for meningitis and encephalitis are commercially available from a limited number of manufacturers at the time of this writing. The few commercial assays generally target the most common viral causes of meningitis and encephalitis that have been conventionally detected by molecular tests (e.g., enterovirus, herpes simplex virus) as well as the most common bacterial causes of meningitis. Note that viruses that traditionally have been identified by IgM detection in CSF (e.g., West Nile virus, deer tick virus) have not been included on these panels due to the limited utility of PCR in these infections. These tests have shown contentious clinical value, primarily due to specificity concerns for some of the viral targets. Bacterial cultures from CSF are still required in conjunction with these assays as there are other bacteria and fungi that can gain access to the CSF that are not conventionally considered in the differential diagnosis for meningitis, and the organism AST results are still needed to guide therapy. Some targets that have been included on these panels may also be inappropriate to test in all populations, such as human parechovirus and *E. coli* K1 which are almost exclusively pediatric causes of meningitis. Results should be interpreted with caution when the patient demographic and the target detected are incongruent.

Diagnostic Mycobacteriology Methods

Mycobacteria encompass many important pathogens in the *Mycobacterium* genus, most notably *Mycobacterium tuberculosis*, *M. avium* complex, and *M. leprae*. Mycobacteria are aerobic bacterial rods which possess higher levels of mycolic acids in their cell walls than most other bacteria. Mycolic acids, alongside other fatty acids such as waxes and phospholipids, render the cell wall of mycobacteria highly hydrophobic. This highly complex lipid content prevents staining of the bacilli by the usual Gram stain method. Special procedures are required to penetrate and stain the cell wall of mycobacteria to visualize the bacteria. This complex lipid wall also protects the mycobacteria from environmental stressors; mycobacteria can easily survive for weeks to months on objects if not exposed to ultraviolet light.

Some mycobacteria are commensal and colonize areas of the body. However, most mycobacteria are obligate or opportunistic pathogens; their presence is considered abnormal and disease-causing. *M. tuberculosis*, which is an organism complex including *M. tuberculosis*, *M. bovis*, the bacillus Calmette-Guérin (BCG) strain and others, causes chronic lung and disseminated disease throughout the world and is highly contagious. Airborne particles of *M. tuberculosis* are generated by persons with pulmonary tuberculosis when they cough, and these particles can remain suspended in air for several hours. *M. leprae* damages peripheral nerves, leading to skin lesions and other complications from loss of sensation of touch. Non-tuberculous mycobacteria (NTM) are comprised of a large group of bacilli, many of which cause disease. NTM include *M. avium* complex, rapidly growing mycobacteria (e.g., *M. abscessus*, *M. fortuitum*, *M. chelonae*), *M. chimaera*, and others. There have been recent reports of invasive infections in open-heart surgery patients due to heater-cooler devices for extracorporeal circulation which were contaminated with *M. chimaera*.

Mycobacteria grow more slowly than other bacteria, with incubation times measured in weeks instead of days. Special media are used for culture. Diagnosis of mycobacterial disease can be achieved using a variety of methods, many of which are used in combination to determine presence and extent of disease.

Specimen Processing

Transmission risk of *M. tuberculosis* is high when collecting a respiratory specimen for mycobacterial studies. Healthcare personnel wear respirators and other protective gear when collecting sputum from a person from whom tuberculosis is suspected. The patient with suspected tuberculosis is placed in a containment room with unidirectional airflow leading into—rather than out of—the room (i.e., negative pressure rooms). Likewise, laboratory workers must protect themselves from potential mycobacteria in the specimen. Biosafety level (BSL) requirements for handling of specimens for mycobacteriology studies are more stringent than that required for specimen handling in the bacteriology laboratory. Stringent containment and laboratory safety practices are essential when performing mycobacterial work. BSL-3 facilities are preferred, but BSL-2 practices (with additional modifications such as negative pressure rooms and additional personal protective equipment) are acceptable. All specimen processing (and subsequent culture work) must take place within a biosafety cabinet.

Due to the slow growth of mycobacteria and their requirement for specific nutrients for growth, mycobacteria can be overgrown by other bacteria or fungi present within those specimens. Thus, mycobacterial specimens collected from non-sterile sources (such as sputum) must be decontaminated before plating occurs. Mucin within respiratory secretions trap mycobacteria; thus, the specimen must be "digested" before decontamination to break apart the mucin to allow for more accurate culture results.

Examination of Direct Specimens

Smear microscopy provides a rapid result of the presence or absence of mycobacteria within the specimen. Recommendations for sputum collection for rule out of *M. tuberculosis* include three smears although rapid nucleic acid testing may now replace one of the smears, for a practical but accurate approach to ruling out disease and discontinuing respiratory isolation. The acid-fast Kinyoun or Ziehl-Neelsen (Z-N) stains based on carbol fuchsin may be used for diagnosis. The Z-N stain involves heating the specimen to break open the tough cell walls of mycobacteria to allow stain penetration, while the Kinyoun stain is a "cold" method, whereby chemicals are used to penetrate the cell wall. Mycobacteria appear characteristically red on the acid-fast stain (Fig. 1.9). Fluorophore stains (e.g., auramine rhodamine, auramine O) are more often routinely used in clinical microbiology laboratories than acid-fast stains for direct specimen examination since the technique is up to ten-fold more sensitive than the carbol fuchsin-based method (Fig. 1.10).

Culture Methods

Culture is the mainstay of diagnosis for mycobacteria, except for *M. leprae* which is not cultivatable in the clinical microbiology laboratory. Egg-based solid Löwenstein-Jensen medium and agar-based solid Middlebrook medium are commonly used in laboratories (Fig. 1.11). Supplements are often added to media to suppress growth of non-mycobacterial organisms. The species name of *M. haemophilum* stands for "heme-loving"; this organism will grow on media which is supplemented with hemin or hemoglobin. Furthermore, this species requires lower incubation temperatures of 30 °C to grow, rather than the typical 35–37 °C at which mycobacterial specimens for culture are usually incubated.

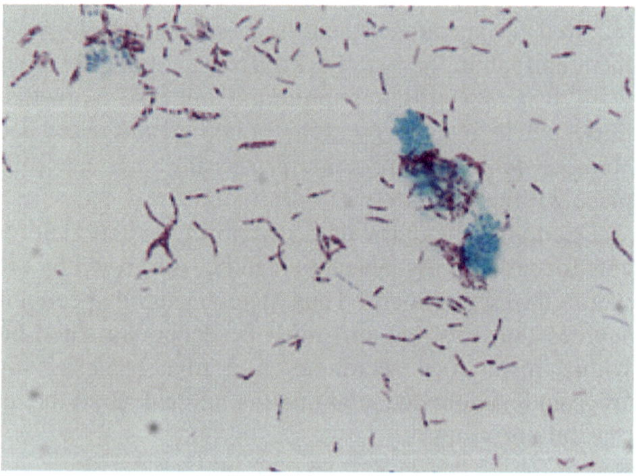

Fig. 1.9 Many beaded mycobacteria positive on direct sputum acid-fast Kinyoun stain (1000× magnification)

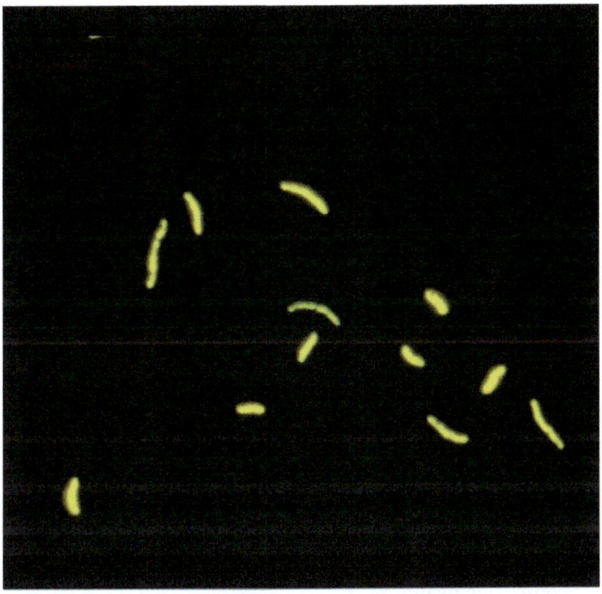

Fig. 1.10 Several mycobacteria staining positive by fluorescence on direct sputum (1000× magnification)

Fig. 1.11 Rough and buff-colored *Mycobacterium tuberculosis* colonies growing on Middlebrook agar

Mycobacterial cultures are incubated for 6–8 weeks. Use of liquid broth medium allows for more rapid growth (sometimes within days) as compared to solid media. Some mycobacteria, most notably *M. tuberculosis*, form microscopic serpentine cords in which bacilli cluster together in parallel chains (Fig. 1.12). Cording is appreciated after organisms have grown in culture and can be seen on staining of the bacilli in liquid media. This phenomenon is a virulence factor and is due to the presence of the glycolipid trehalose dimycolate. Rapidly growing mycobacteria are those organisms which grow within 5–7 days of subculture on solid media. Slowly growing mycobacteria have traditionally been classified using the Runyon

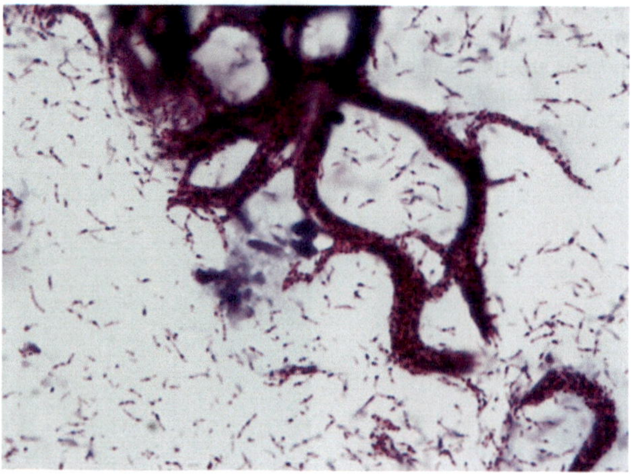

Fig. 1.12 Cording of *Mycobacterium tuberculosis* in liquid culture broth (Kinyoun stain, 1000× magnification)

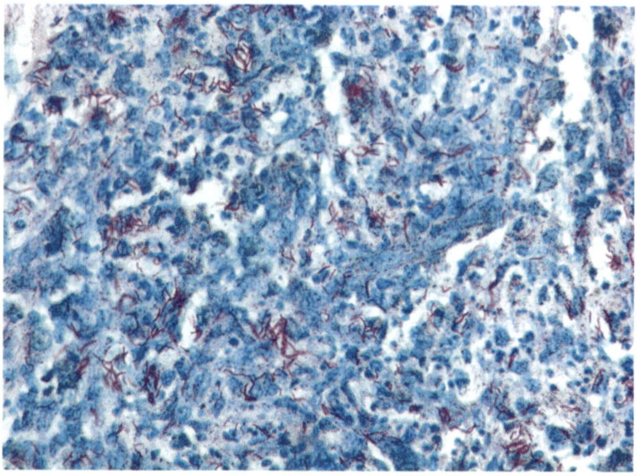

Fig. 1.13 Histopathology of acid-fast stain of necrotizing cheek lesion demonstrating many mycobacteria; *Mycobacterium kansasii* was recovered in culture (1000× magnification)

grouping, according to whether colony pigment develops and, if so, whether it develops when colonies are grown under light or dark conditions. Organisms were divided into nonchromogens, photochromogens (i.e., colonies grown in light develop pigment but those grown in dark are not pigmented), and scotochromogens (i.e., develop pigmentation in the light and in the dark). However, this classification is outmoded, and more accurate diagnostic methods are now used.

Histopathology

Mycobacteria appear on histopathology in acid-fast-stained formalin-fixed paraffin-embedded (FFPE) sections or on cytologic examination as acid-fast positive bacilli (Fig. 1.13). They cannot be seen on the hematoxylin and eosin (H&E) stain. The bacilli may appear beaded or banded on acid-fast stain. Mycobacterial species

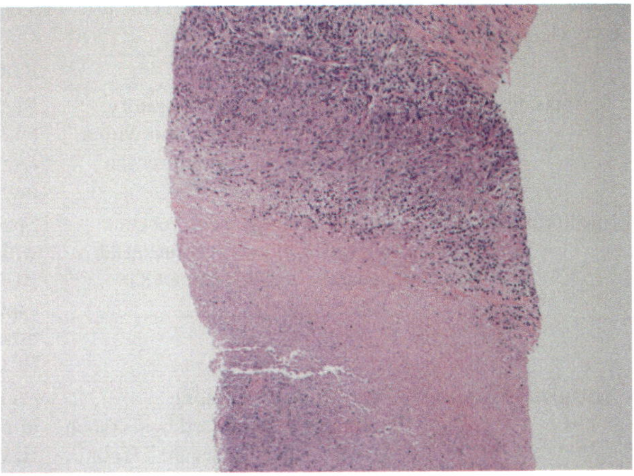

Fig. 1.14 Mediastinal mass with necrotizing granulomas; *Mycobacterium avium* complex was recovered in culture (100× magnification)

cannot be determined by histopathology alone; culture is required for identification. A necrotizing or non-necrotizing granulomatous inflammatory reaction consisting of macrophages accompanies mycobacteria (Fig. 1.14). The bacillary load of mycobacteria can be extremely low; therefore, slides must be carefully examined, as even the presence of a single acid-fast-positive bacillus can be significant.

Immunodiagnostic Tests of TB Infection

Determining if a patient has immunologic evidence of TB infection—"germs in the body"—also contributes to the diagnosis of tuberculosis disease, especially in those cases when an organism cannot be detected directly. Two tests are available to determine if an individual is infected with MTB: the tuberculin skin test (TST) and the interferon (INF)-γ release assays (IGRAs) (Table 1.1). These test results are interpreted the same for children as for adults. Both types of tests produce continuous results, but the tests are interpreted in a binary fashion with cutoff values used to interpret results as "positive" or "negative." The definitive TST uses five tuberculin units of purified protein derivative (PPD) stabilized in Tween 80. A 26- or 27-gauge needle and a graduated syringe are used to inject 0.1 mL of PPD intradermally into the volar surface of the forearm; the immediate appearance of a wheal indicates correct technique. Delayed hypersensitivity reaction to TST usually peaks at 48–72 h. In some individuals, reaction occurs after 72 h and should be measured at that time. The diameter of induration, not erythema, is measured perpendicular to the axis of administration and recorded in millimeters. TST result should not be recorded as simply "positive" or "negative." Of note, it may take up to 10 weeks after infection occurs for an individual to react to the TST.

A nonreactive TST result does not exclude MTB infection or disease, as a variety of factors can lower tuberculin reactivity. Approximately 20% of immunocompetent children with culture-confirmed TB disease do not react initially to the TST; the rate is even higher in individuals that are significantly immunocompromised as a result

Table 1.1 Comparison of the tuberculin skin test (TST) and interferon-gamma release assay (IGRA)

	TST	IGRA
Test setting	Does not require a laboratory; requires a follow-up visit with a trained person to interpret the result in 48–47 h	Requires a blood draw and laboratory equipment; does not require a follow-up visit for result interpretation
Specificity issues	False positives due to recent nontuberculous mycobacterial infection and/or prior BCG vaccination	Specificity higher than that of TST, with no false positives due to prior BCG vaccination; specificity, especially in settings of low tuberculosis burden and among BCG-vaccinated individuals
Sensitivity issues	Poor sensitivity among immunocompromised hosts (such as HIV or active, severe TB) or with waning of immunity	Poor sensitivity among immunocompromised hosts (such as HIV or active, severe TB) or with waning of immunity
Ability to differentiate TB infection from disease	No	No

BCG bacillus Calmette-Guerin, *HIV* human immunodeficiency virus, *TB* tuberculosis

of disease or medication. Improper storage, dilution, placement, and interpretation of the TST can cause false-negative results.

The most significant causes of false-positive TST reactions are recent nontuberculous mycobacterial (NTM) infection and prior BCG vaccination. NTM infection, which occurs more frequently near the equator, usually causes a cross-reaction of ≤10 mm (but can be larger); cross-reactivity can last for several months. In studies of BCG-vaccinated newborns, only 50% have a positive TST result, and 80% to 90% lose such reactivity within 5 years. Older children or adults have higher initial and longer responses to BCG, but most lose tuberculin reactivity within 10 years of vaccination. The degree of reactivity also is affected by BCG product and nutritional status. Of note, countries that use BCG vaccine frequently have high rates of endemic TB, and studies have demonstrated that a positive TST in a previously BCG-vaccinated child who is a close contact with an active TB case likely indicates MTB infection.

Three different cutoff values are used to interpret TST reactivity. These cutoff values represent a statistical attempt to minimize false-positive or false-negative readings and vary according to individual and epidemiologic factors, of which recent exposure to MTB is the most heavily weighted. For children at highest risk of infection progressing to disease, induration diameter of ≥5 mm is classified as a positive result. For other high-risk groups, an induration diameter ≥10 mm is a positive result. For low-risk children, an induration diameter ≥15 mm is a positive result.

There are two commercially available IGRAs: QuantiFERON®-TB Gold (QFT; Cellestis/Qiagen, Carnegie, Australia) and T-SPOT®.TB (T-SPOT; Oxford Immunotec, Abingdon, U.K.). In terms of performance, neither test is preferred over the other. IGRAs measure INF-γ secreted by the patient's T-lymphocytes (QFT) or the number of IFN-γ secreting lymphocytes (T-SPOT) upon *ex vivo* stimulation with *M. tuberculosis*-specific antigens that are not found in BCG vaccine strains or most NTM species (except *M. marinum, M. kansasii, M. szulgai,* and *M. flavescens*). Both IGRAs utilize positive and negative controls; if either control fails, the result is deemed indeterminate (QFT) or invalid (T-SPOT). For T-SPOT only, an invalid result is classified as borderline. Unlike the TST, each IGRA has only one cutoff value regardless of the patient's exposure history or immune status. However, some experts have questioned this lack of risk stratification and suggest a need for further refinement of IGRA cutoff values.

Studies have demonstrated that IGRAs have a higher specificity than the TST for tuberculosis infection, particularly in settings of low tuberculosis burden and among BCG-vaccinated children (Table 1.1). One meta-analysis estimated a specificity of 85% to 95% for IGRAs in BCG-vaccinated individuals, compared to 45% to 60% for the TST. IGRAs and the TST have comparable sensitivities in immunocompetent individuals. However, like the TST, IGRAs have poor sensitivity among immunocompromised hosts and those with severe tuberculosis disease and cannot differentiate TB infection from disease.

MALDI-TOF Mass Spectrometry

Matrix-assisted laser desorption ionization-time-of-flight mass spectrometry (MALDI-TOF MS) is a form of ionization and mass spectrometry that has revolutionized bacterial identification in clinical laboratories. For MALDI-TOF MS identification, isolated bacterial colonies are mixed with matrix, typically an acid, and applied to a metal plate that is then irradiated with a laser to ionize the bacterial proteins and remove them from the plate (desorption). The proteins and peptides are then accelerated toward a detector; the time that it takes for an ionized particle to reach the detector is determined by its mass-to-charge ratio. For a given bacterium, this will generate a spectrum of protein peaks that can then be compared to a database and matched with known bacteria. However, the use of MALDI-TOF MS for mycobacteria identification has lagged behind its utilization for other bacteria. Only recently in 2017 did VITEK MS (bioMérieux, Durham, NC), one of the two commercially available MALDI-TOF MS systems in the US, receive FDA clearance for the identification of 19 species of *Mycobacterium*, include *Mycobacterium tuberculosis* complex, *Mycobacterium avium*, and *Mycobacterium abscessus*. At of the time of writing, the other system, MALDI Biotyper (Bruker Daltronics, Billerica, MA), has mycobacterial identification limited to their research-use only database in the United States, but has an extensive library approved for *in vitro* diagnosis available in other countries.

Unlike other bacteria and yeasts, the use of MALDI-TOF MS for mycobacteria requires additional processing prior to analysis due to the nature of their cell walls. For both the VITEK MS and MALDI Biotyper, mycobacteria are inactivated and protein extracted. The extracted protein solution can then be applied to the MALDI target plate. In addition to identification from solid media, some laboratories have developed protocols to identify mycobacteria directly from liquid media, which can reduce the time to identification by several weeks. Due to the pathogenicity of MTB, the extraction and inactivation should be performed in a BSL-3 laboratory or a BSL-2 laboratory using BSL-3 practices. It is critical that the mycobacteria are killed during the inactivation or extraction step, as viable mycobacteria on the target pose a biosafety risk.

AFB Identification by Sequencing

The gold standard for MTB and non-tuberculous mycobacterial identification is nucleic acid amplification and sequencing. This has largely replaced biochemical phenotypic testing, which is time-consuming and can provide ambiguous results, and HPLC analysis of mycolic acid, which often is unable to differentiate closely related species and is technically challenging. Colonies of mycobacteria grown on solid media are collected and nucleic acid is extracted, often after a heat inactivation step. Cell lysis is typically performed mechanically, such as with bead-beating or sonication, due to the resilience of the mycobacterial cell wall. Nucleic acid can be separated from the cell lysate using magnetic beads, spin columns, or traditional methods like phenol/chloroform extraction with ethanol precipitation. This final purification step is critical to generate a good template for subsequent amplification and sequencing.

Genes that are commonly amplified and sequenced include 16S rRNA, *hsp65*, and *rpoB*. All three are housekeeping genes that are conserved across all mycobacteria though the latter two genes exhibit more variability across mycobacteria species and offer higher resolution. For all three targets, conserved regions are targeted with oligonucleotides to allow amplification of variable regions. This is generally performed for short reads of 500–800 base pairs, but longer reads can be used to further distinguish closely related species. Amplified PCR products are then sequenced using Sanger sequencing or pyrosequencing, and the results are compared to sequences of known species in a database. There are many databases available, depending on the gene, including publicly available databases like NCBI GenBank.

Beyond using sequencing for identification of individual isolates, technology such as whole genome sequencing (WGS) are being utilized in public health settings for strain typing of *M. tuberculosis* complex. Unlike sequencing of individual genes for identification, WGS and other stain typing assays offer much higher discrimination to identify epidemiologic links between cases of tuberculosis and assist tuberculosis control programs. WGS also has the potential in clinical laboratories to identify mutations associated with antituberculosis drug resistance. Genotypic resistance can offer results faster than phenotypic susceptibility testing, which can take 2–3 weeks after an isolate has been identified.

Antimicrobial Susceptibility Testing

When an isolate of *M. tuberculosis* is cultivated, antimicrobial susceptibility testing (AST) should be performed. Multidrug resistance is common, and treatment is a long-term process. Susceptibility testing of NTM should be performed on clinically significant isolates.

Summary

Mycobacteria differ from other bacteria in the high mycolic acid content of their cell walls. Acid-fast stains can highlight these organisms. Mycobacteria doubling time is very long; therefore, cultures must be incubated for weeks. Safety precautions must be followed when working with mycobacterial cultures in the laboratory. Cultivated organisms can be identified using MALDI-TOF MS or sequencing, and several platforms are available for testing. Nucleic acid amplification and sequencing is also available from clinical specimens for rapid and accurate diagnosis of certain mycobacterial diseases. Diagnostic tools beyond culture include immunodiagnostics and whole genome sequencing.

Case 1

Angela Ma and Marc Roger Couturier

A previously healthy 21-year-old male with no recent travel presented at the emergency department with 5-days history of fever, chills, abdominal pain, and diarrhea. He decided to seek medical attention upon his symptoms not resolving within 48 h and development of blood and mucous in the stool. One week prior, the patient had dined out with his roommate and eaten a grilled chicken burger. His roommate had ordered pizza and was asymptomatic thus far. After initial consultation with the physician, the patient was given a vial of enteric transport media and instructed to submit a stool sample for microbiological testing at his earliest convenience. Considering the duration of the patient's symptoms, the physician ordered both a stool culture and a multiplex molecular panel for gastrointestinal (GI) pathogens. The stool culture could recover common bacterial agents of acute gastroenteritis including *Campylobacter* spp., *Salmonella*, *Shigella* spp., Shiga-like toxin-producing *Escherichia coli*, *Yersinia enterocolitica*, *Vibrio* spp., *Aeromonas* spp., and *Plesiomonas shigelloides* if viable cells were present in the sample, and the multiplex molecular test would detect nucleic acids of viral and parasitic pathogens in addition to most common GI-pathogenic bacteria.

The patient was administered oral rehydration therapy while in the emergency department and within 3 h, he was diagnosed with *Campylobacter* gastroenteritis by nucleic acid detection. Following consultation with the care team, it was decided that empiric antimicrobial therapy would not be necessary and the patient was discharged with recommendations to stay hydrated. After 48 h incubation from inoculation of the stool sample onto primary cultures, spready, gray colonies were observed on the Campy CVA agar plates incubated in a microaerobic atmosphere at 42 °C (Fig. 1.15). Gram stains of the colonies showed Gram-negative, "seagull-shaped" rods that correlated with *Campylobacter* spp. (Fig. 1.16). Oxidase was positive, catalase was negative, and hippurate hydrolysis was positive, suggesting *C. jejuni* subsp. *doylei*. Subcultures of the colonies were confirmed as *C. jejuni* by matrix-assisted laser desorption ionization-time-of-flight (MALDI-TOF) mass spectrometry (subspecies identification was not possible). The patient fully recovered 3 days after discharge and no antimicrobial therapy was necessary.

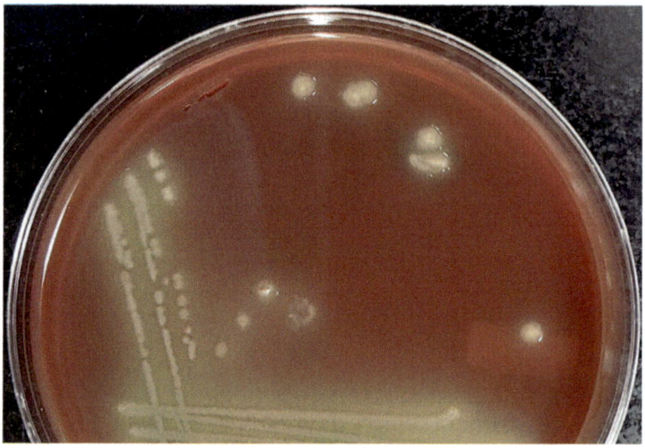

Fig. 1.15 *C. jejuni* on CVA plates from primary stool cultures

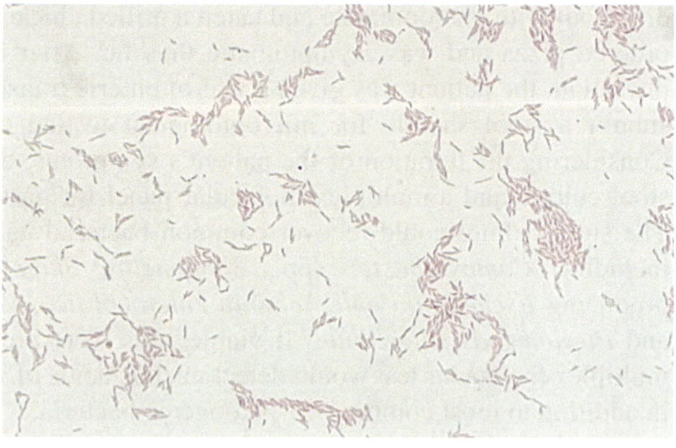

Fig. 1.16 Gram stain of *C. jejuni* with distinctive "seagull-shaped" Gram-negative rods captured at 1000× magnification

Discussion

Campylobacter spp. are non-spore forming, Gram-negative, spiral-shaped bacilli that are motile via an amphitrichous flagellum. Common descriptors of *Campylobacter* spp. Gram stain morphology include "S-shaped" or "seagull wings." Most *Campylobacter* species are fastidious and require a neutral pH and a microaerobic atmosphere comprised of decreased O_2 and increased CO_2 concentrations at 5% and 10%, respectively, for optimal growth. The "thermophilic" *Campylobacter* species most commonly recovered from humans (*C. jejuni, C. coli, C. upsaliensis,* and *C. lari*) are capable of growth at 42 °C, allowing them to outgrow other contaminating enteric microaerobic flora. Some species of *Campylobacter* such as *C. butzleri* and *C. concisus* can grow aerobically or anaerobically, respectively.

Bacterial infection from the *Campylobacter* genus is the leading cause of acute gastrointestinal illness in the United States. *C. jejuni* subsp. *jejuni* and *C. coli* are the most common etiologic agents for campylobacteriosis, and it is estimated that *C. jejuni* infection affects over 1.3 million people annually. However, the actual incidence of *Campylobacter* infection worldwide is likely significantly greater than current reports due to underdiagnosis and underreporting. *Campylobacter* spp. are zoonotic organisms with many of its primary reservoirs being livestock and domesticated animals including poultry, cattle, sheep, pigs, canines, and felines. Approximately 68% of poultry is contaminated with *Campylobacter* spp. at the farm and retail level. However, various household pets including reptiles have been known to harbor *Campylobacter* spp. as well. The main route of transmission of *C. jejuni* is fecal-oral and is commonly associated with ingestion of improperly handled or undercooked food. Undercooked or raw poultry meat is the leading cause of campylobacteriosis, but infection has also been reported from cases of ingestion of contaminated water or unpasteurized milk. Direct person-to-person or contact with animals can also transmit *Campylobacter* spp. Historically, subtherapeutic levels of antibiotics including β-lactams, fluoroquinolones, and tetracyclines were regularly administered to livestock in efforts to promote growth and combat spread of disease. Such practices however, resulted in the selection of antibiotic-resistant *C. jejuni* strains, particularly in poultry and subsequent corresponding increases in antibiotic resistance were also observed in human infections. Use of antibiotic growth promoters in poultry has since been banned in the European Union and the United States since 2006 and 2017, respectively.

The majority of *C. jejuni* infections are sporadic although outbreaks are rare, they occur more often in the spring and fall months. Ingestion of less than 500 organisms can cause campylobacteriosis. The incubation period of *C. jejuni* ranges from 2 to 4 days on average and the disease is marked by general symptoms of acute gastrointestinal illness including diarrhea, abdominal pain, fever, and sometimes vomiting. *C. jejuni* facilitates human diarrheal disease through adhesion and invasion of the colonic epithelium and stimulation of a reactionary inflammatory response. Clinically, campylobacteriosis is indistinguishable from other invasive foodborne bacterial pathogens such as *Salmonella*, *Shigella* spp., and *Yersinia enterocolitica*.

Treatment for campylobacteriosis is typically unnecessary since the disease is self-limiting and resolves within a week of onset. Hydration and electrolyte therapy are sufficient in most cases of infection. Antibiotic treatment should only be considered if a high fever persists, dysentery is noted, prolonged illness or relapsing infection, or if the patient is pregnant or immunocompromised. Complications due to *Campylobacter* infection are rare but can occur. Campylobacteriosis can mimic acute appendicitis and potentially lead to unnecessary surgical intervention. *Campylobacter* spp. can cause extraintestinal infections such as bacteremia, hepatitis, pancreatitis, and urinary tract infection. Beyond extraintestinal complications, it is important to recognize the potential for *C. jejuni* infection to lead to post-infection sequelae. Guillain–Barré syndrome (GBS) is an acute paralytic disease of the peripheral nervous system characterized by a progressive weakness in the limbs that can arise following *Campylobacter* infection. A host autoimmune response to sialic acid present on the lipooligosaccharide of *C. jejuni* results in the production of anti-ganglioside antibodies that demyelinate the peripheral nerves. Reactive arthritis and irritable bowel syndrome also affect 2–5% and 9–13% of individuals following *C. jejuni* infection, respectively.

Although *C. jejuni* and *C. coli* are the most prevalent *Campylobacter* species diagnosed that cause human infection, other members of the genus can cause disease as well. Examples include *C. fetus* subspecies *fetus* and *C. upsaliensis*. Both organisms can cause diarrheal illness but *C. fetus* subsp. *fetus* is more commonly associated with bacteremia and systemic disease in immunocompromised individuals. Culture and incubation conditions for the isolation of *C. fetus* subsp. *fetus* and *C. upsaliensis* are different from those routinely used for *C. jejuni* and *C. coli* and may be missed if not considered in the differential diagnosis. *C. fetus* subsp. *fetus* requires an incubation temperature of 25 °C whereas primary isolation of *C. upsaliensis* is improved with the use of selective, charcoal-based media.

Stool is the optimal specimen type for the isolation of *Campylobacter* spp. from cases of acute gastroenteritis. Blood, tissue, and body fluids are also acceptable for culture when *Campylobacter* spp. is considered a potential etiologic agent in non-gastrointestinal infection. Routine Gram stains are not typically performed directly from stool specimens; however, *Campylobacter* spp. do stain better with carbol-fuchsin or 0.1% aqueous basic fuchsin counterstains as opposed to safranin that is typically used. This modified Gram stain has been reported to allow direct detection of *Campylobacter* spp. from stool specimens or pure culture although the method is infrequently used. If *Campylobacter* is suspected from a primary Gram stain but the organism is very pale and difficult to visualize, a repeat Gram stain where the safranin is left on for 2–5 min can increase stain intensity. To increase the likelihood of isolating *Campylobacter* spp., selective media such as the Campy CVA, containing additional nutrients and antimicrobial agents are often utilized. Plates are typically incubated at 42 °C in a microaerobic environment for at least 48 h before examining for colonies resembling *Campylobacter* spp. Microaerobic conditions can be generated through the inclusion of gas generator packs or evacuation and replacement of gas in larger jars or chambers. Colony morphology of *Campylobacter* spp. is highly variable, colonies can range from flat, watery, and spready to small, convex, and round and be gray to yellow in color (Fig. 1.17). *Campylobacter* spp. are oxidase positive

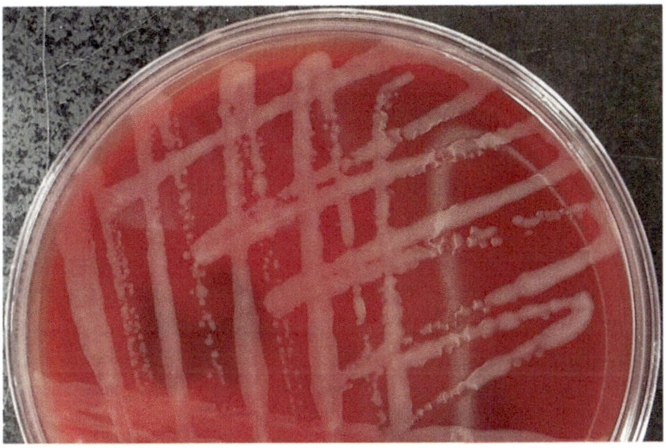

Fig. 1.17 Pure culture of *C. jejuni* on sheep blood agar

and *C. jejuni* subsp. *jejuni* is hippurate hydrolysis positive while other commonly encountered *Campylobacter* species are variable (*C. jejuni* subsp. *doylei)* or negative. Colonies suspected to be *Campylobacter* spp. can be identified by MALDI-TOF mass spectrometry or 16S rRNA gene sequencing though the standard 500 base pair 5′ region of the 16S rRNA gene does not provide adequate separation of most species.

Culture-independent diagnostic methods from fecal specimens include antigen testing and nucleic acid amplification tests. Enzyme immunoassays detect *Campylobacter*-specific antigens present in various *Campylobacter* spp. but typically suffer from poor sensitivity and positive predictive value and thus require confirmation by culture or detection of nucleic acid. A number of FDA-cleared multiplex platforms for the detection of nucleic acid from common gastrointestinal pathogens are currently available for commercial use, and some cover viral and parasitic targets in addition to bacterial targets.

Antimicrobial susceptibility testing guidelines and interpretations are available for *C. jejuni* and *C. coli* by the Clinical and Laboratory Standards Institute (CLSI). Antimicrobial agents recommended for testing include ciprofloxacin, erythromycin, tetracycline, and/or doxycycline by broth microdilution or disk diffusion methods. Macrolides and fluoroquinolones are the drugs of choice for antimicrobial therapy in cases of prolonged infection but resistance to both classes have been reported, particularly in Asia and Africa. Furthermore, *Campylobacter* spp. possess intrinsic resistance to some β-lactams, polymyxin/colistin, and trimethoprim, sulfamethoxazole.

Summary Points

- *Campylobacter* spp. are generally curved, Gram-negative rods of zoonotic origin that can cause acute gastroenteritis and extraintestinal disease.
- Sequelae associated with *Campylobacter* infection include Guillain–Barré syndrome, reactive arthritis, and irritable bowel syndrome.

- *Campylobacter* isolation by culture requires a microaerobic atmosphere, selective solid media or filtration of stool, and extended incubation for a minimum of 72 h, often at 42 °C.
- Antimicrobial therapy is typically unnecessary in cases of uncomplicated *Campylobacter* infection. If treatment is required, quinolone and macrolide agents are recommended but increasing rates of acquired resistance to erythromycin and ciprofloxacin have been associated with historical use in animals, particularly poultry.
- Detection of *Campylobacter* spp. by culture-independent diagnostic tests such as antigen detection and PCR have become more common and provide much faster results than culture.

Selected References

Dasti JI, Tareen AM, Lugert R, Zautner AE, Gross U. *Campylobacter jejuni*: a brief overview on pathogenicity-associated factors and disease-mediating mechanisms. Int J Med Microbiol. 2010;300(4):205–11.

Friedman CR, Hoekstra RM, Samuel M, Marcus R, Bender J, Shiferaw B, Reddy S, Ahuja SD, Helfrick DL, Hardnett F, Carter M, Anderson B, Tauxe RV; Emerging Infections Program FoodNet Working Group. Risk factors for sporadic *Campylobacter* infection in the United States: a case-control study in FoodNet sites. Clin Infect Dis. 2004;38 Suppl 3:S285–96.

Geissler AL, Bustos Carrillo F, Swanson K, Patrick ME, Fullerton KE, Bennett C, Barrett K, Mahon BE. Increasing *Campylobacter* infections, outbreaks, and antimicrobial resistance in the United States, 2004–2012. Clin Infect Dis. 2017;65(10):1624–31.

Hsieh YH, Simpson S, Kerdahi K, Sulaiman IM. A comparative evaluation study of growth conditions for culturing the isolates of *Campylobacter* spp. Curr Microbiol. 2018;75(1):71–8.

Case 2

Megan H. Amerson-Brown and James J. Dunn

A 3-year-old previously healthy female presented to the emergency department after 3 days of illness that consisted of fever, headache, lethargy, and body aches. Two purpuric lesions were noted on her left arm. Laboratory testing revealed a mildly elevated white blood cell count (12,300/μL) with 46% neutrophils, 11% bands, 41% lymphocytes, and 2% monocytes. Microscopic examination of the Wright-Giemsa stained peripheral blood smear showed a number of neutrophils that contained intracellular, coccoid structures consistent with the presence of bacteria (Fig. 1.18). Blood cultures were collected.

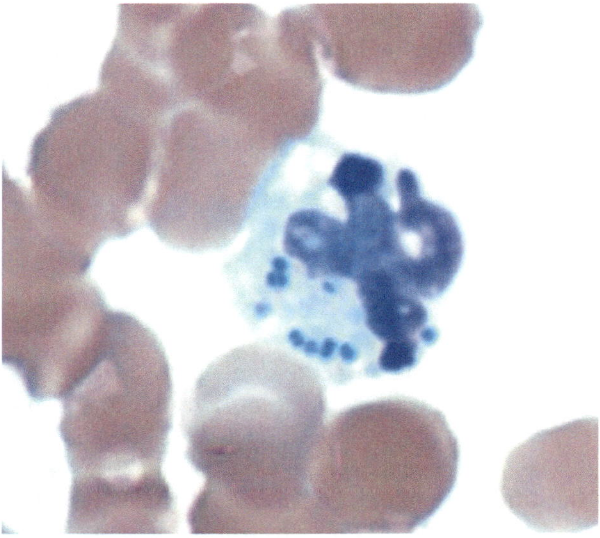

Fig. 1.18 Peripheral blood smear with intracellular coccoid-like organisms. (Wright-Giemsa stain, magnification 1000×)

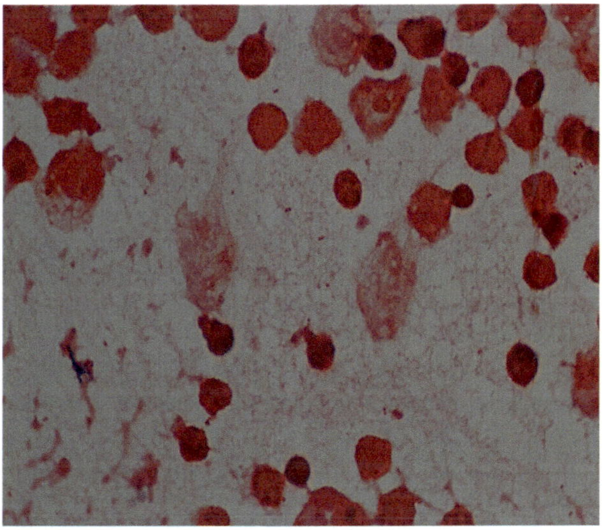

Fig. 1.19 Gram stain of CSF with the presence of Gram-negative diplococci (magnification 1000×)

The patient had abnormal coagulation studies with INR = 2.2 (reference range, 0.8–1.2), PTT = 196 s (reference range, 25.6–32.4 s) and D-dimer >20 µg/mL (reference range, ≤0.4 µg/mL) suggesting disseminated intravascular coagulation. A CSF was obtained by lumbar puncture and showed an elevated white blood cell count of 18,770/µL (reference range, 0–3/µL) with 79% neutrophils, 10% lymphocytes and 11% monocytes. The CSF red blood cell count was 110/µL (reference range, 0/µL), glucose was undetectable (<20 mg/dL) (normal range is 50–70% of serum glucose concentration), and protein was 491 mg/dL (normal range, 15–45 mg/dL). A Gram stain of the CSF revealed Gram-negative diplococci suggestive of meningococcal infection (Fig. 1.19). At this point, the patient was admitted to the hospital and started on IV ceftriaxone. Over the next day, more purpuric skin lesions appeared and the patient became acidotic with poor perfusion and azotemia. The blood culture was positive after 12 h of incubation and the Gram stain showed the same Gram-negative diplococci observed in the CSF. The isolate from cultures of both blood and CSF were identified by MALDI-TOF mass spectrometry as *Neisseria meningitidis*.

Discussion

N. meningitidis is an aerobic, nonmotile Gram-negative coccus that classically occurs in pairs with the adjacent sides flattened giving it a diplococci appearance. The organism produces catalase and oxidase and may sometimes be encapsulated. *N. meningitidis* oxidizes glucose and maltose which distinguishes it from other Neisseria species like *N. gonorrhoeae* which oxidizes only glucose. Meningococci are divided into 12 serogroups with serogroups A, B, C, W-135, X and Y being the most common cause of invasive disease such as bacteremia and meningitis worldwide.

Humans are the only natural reservoir for *N. meningitidis*. Intermittent, asymptomatic colonization of the nasopharynx (NP) with *N. meningitidis* can occur. The prevalence of colonization ranges from about 2% in young children to approximately 25% in adolescents and young adults. Transmission occurs person to person via respiratory droplets. How the organism converts from NP carriage to invasive disease is not well understood. But there are several factors that are associated with invasive disease including you age, crowding, certain immune deficiencies, and asplenia. Sporadic infections of meningococcal disease typically occur in children younger than 1 year. However, in outbreak situations, older children and young adults are more often affected.

Clinical manifestations of *N. meningitidis* infection most commonly include meningitis (50% of cases), bloodstream infection (meningococcemia) (about 35% of cases), or both concurrently (5% of cases). The onset of symptoms associated with bloodstream infection can be abrupt, with rapid onset of fever, chills, malaise, and lethargy. Petechial or purpuric skin lesions can be seen in most patients with meningococcemia due to activation of host inflammatory factors. Disseminated intravascular coagulation can occur due to activation of the host coagulation system by endotoxin produced by the organism. Pneumonia caused by *N. meningitidis* usually occurs in conjunction with meningococcemia or meningitis and accounts for about 10% of cases of infection. Other clinical syndromes associated with meningococcal infection include conjunctivitis, septic arthritis, pericarditis and myocarditis. The overall mortality rate of invasive meningococcal disease can be as high as 15–20% and long-term complications can occur in 20–50% of patients. Long-term sequelae of meningococcal meningitis include hearing loss, cognitive difficulties, and visual disturbances. Patients with *N. meningitidis* septicemia may experience skin scarring and limb amputation.

Culture of blood and/or CSF is considered the gold standard for diagnosis of invasive meningococcal infection. Gram stain performed directly on CSF with the presence of Gram-negative diplococci (Fig. 1.19) can be a rapid indication of the possibility of meningococcal meningitis. In fulminant infection, the organism may also be seen in direct stains of peripheral blood as with the patient in this case (Fig. 1.18). Importantly, blood and CSF specimens should be collected prior to the initiation of antimicrobial therapy as *N. meningitidis* can be rendered nonviable in 15–30 min once antibiotics have been administered, thus decreasing the sensitivity of detection by culture methods. The organism can grow on sheep blood and chocolate agars incubated at 35 °C in 5% CO_2 and once growing can be readily identified by standard biochemical tests, automated identification systems, or MALDI-TOF mass spectrometry. More recently, nucleic acid amplification tests have been employed to detect *N. meningitidis* directly in CSF and sterile body fluids such as blood cultures and can be useful in those instances when prior antimicrobial therapy precludes isolation in culture.

A number of meningococcal vaccines are currently licensed in many countries for prevention of meningococcal disease. The most common ones are quadrivalent and include serogroups A, C, Y, and W-135. A serogroup B meningococcal vaccine has also been licensed separately. Vaccination is recommended for adolescents and certain at-risk individuals in other age groups. Close contacts of patients with invasive meningococcal disease are at increased risk of infection and should promptly receive antibiotic chemoprophylaxis, even if they have previously received the meningococcal vaccine.

Summary Points

- *N. meningitidis* can cause significant invasive disease such as bacteremia and meningitis alone or concurrently.
- Initial clinical manifestations of meningococcal infection may be insidious and nonspecific but can rapidly progress to fulminant disease with concomitant disseminated intravascular coagulation and infection of multiple organs.
- Detection or recovery of *N. meningitidis* from otherwise sterile body fluids is a significant finding and should be reported promptly.
- Individuals at high risk of meningococcal infection should receive one of the currently available vaccines and close contacts of patients with invasive infection should get antibiotic chemoprophylaxis.

Selected References

Cohn A, MacNeil J. The changing epidemiology of meningococcal disease. Infect Dis Clin North Am. 2015;29:667–77.

Dwilow R, Fanella S. Invasive meningococcal disease in the 21st century—an update for the clinician. Curr Neurol Neurosci Rep. 2015;15:2.
Gianchecchi E, Torelli A, Piccini G, Piccirella S, Montomoli E. Neisseria meningitidis infection: who, when and where? Expert Rev Anti Infect Ther. 2015;13:1249–63.
MacNeil JR, Blain AE, Wang X, Cohn AC. Current epidemiology and trends in meningococcal disease—United States, 1996–2015. Clin Infect Dis. 2018;66:1276–81.

Case 3

Marc Roger Couturier

A 23-year-old graduate student from southeast Idaho presented to the emergency department with a large wound on his elbow from a traumatic injury sustained a week earlier. The patient had been mountain biking and backcountry camping with his friends in the deserts of southern Utah. He lost control of his bike on slick-rock during a monsoon rainstorm and crashed into a large rock formation. The patient was ejected into a rocky, sandy creek bed where he took most of the impact on his bare elbow. He suffered several abrasions to the elbow which were rinsed with tap water from his biking bottle and cleansed with alcohol-based hand sanitizer. The patient was unable to clean the wound with soap and warm water until he returned from the trip, approximately 18–24 h later.

Two days before seeking medical care (3 days after the trauma), the patient noted yellowish discharge from the wound and increasing warmth and redness. His body temperature was self-recorded at a maximum of 100.4 °F. At presentation to the emergency department, the wound showed mild, localized cellulitis and a small cutaneous abscess with purulent drainage. His body temperature was 101.1 °F. The emergency physicians conducted an incision and drainage, lavage, and limited debridement of the superficial tissue at the wound site. Tissue and abscess drainage specimens were collected for cultures including: aerobic and anaerobic bacteria, acid-fast bacilli, and fungi. Upon further discussion, the patient revealed that he experienced intermittent and generally moderate flares of irritable bowel syndrome for which he took low-dose prednisone daily (10 mg/day). The patient was current on all vaccines, including tetanus. He was provided a prescription for trimethoprim-sulfamethoxazole taken twice daily for 14 days as empiric therapy, pending culture results. The patient was allowed to return home and instructed to keep the wound clean and dry and to return immediately if the wound continued to worsen and/or the fever persisted or increased.

The Gram stain was negative for organisms from the tissue. The drainage was noted as having beaded, branching Gram-positive rods and few polymorphonuclear leukocytes (Fig. 1.20). Given the soil exposure, aerobic actinomycetes and acid-fast

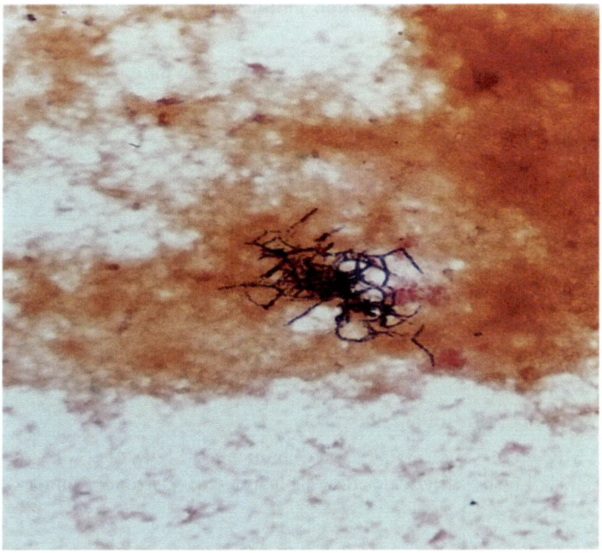

Fig. 1.20 Gram stain of wound drainage showing beaded, branching Gram-positive rods suggestive of *Nocardia* spp.

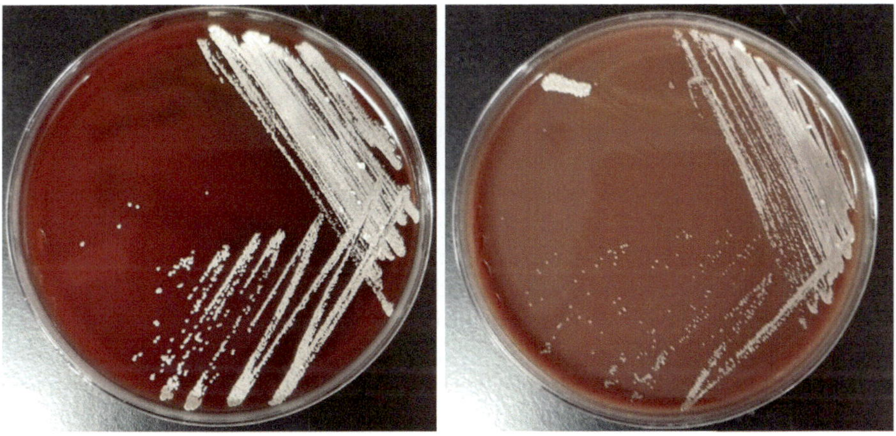

Fig. 1.21 Blood agar (left panel) and chocolate agar (right panel) plates showing characteristic colony growth of *Nocardia brasiliensis*

bacilli were of concern. Growth was reported by the laboratory after 36 h and a preliminary report documented white, chalky, dry colonies resembling *Nocardia* on both blood and chocolate agar (Fig. 1.21). The organism was identified as *Nocardia brasiliensis* by MALDI-TOF mass spectrometry. Antimicrobial susceptibility testing was performed by broth microdilution and revealed predictable sensitivity to trimethoprim-sulfamethoxazole (Table 1.2). The patient completed a 30-day course of antibiotics and experienced uncomplicated wound healing. No further medical care was required.

Table 1.2 Broth microdilution antimicrobial susceptibility testing results for *Nocardia brasiliensis* isolated from the elbow

Antimicrobial	Minimum inhibitory concentration (μg/mL)[a]	Interpretation
Amikacin	2	Susceptible
Amoxicillin clavulanate	1/0.5	Susceptible
Ceftriaxone	2	Susceptible
Ciprofloxacin	**>32**	**Resistant**
Clarithromycin	**>32**	**Resistant**
Doxycycline	0.25	Susceptible
Imipenem	0.125	Susceptible
Linezolid	1	Susceptible
Minocycline	0.125	Susceptible
Tobramycin	1	Susceptible
Trimethoprim/sulfamethoxazole	1/19	Susceptible

[a] MIC cutoffs interpreted according to CLSI M24-A2
The bold values indicates the results that are resistant or abnormal

Discussion

Nocardia belongs to the aerobic actinomycete group, which includes several other common environmental organisms that can be responsible for opportunistic infections of humans (e.g., *Mycobacterium, Streptomyces, Nocardiopsis, Tsukamurella, Actinomadura*). *Nocardia* is the most commonly isolated member of the aerobic actinomycetes from human infections after *Mycobacterium*. The genus *Nocardia* is comprised of Gram-positive rods, which often are visible as beaded and branching forms in direct stains and stains from colonies. The genus is comprised of over 85 species, with many more being described with the advent of whole genome sequencing. Previous species designations may have often included multiple species within a given name, attributed to morphologic variants. Specification of *Nocardia* was historically based on antimicrobial susceptibility patterns; however, this has been shown to be inaccurate and unreliable. *Nocardia* produce aerial hyphae and mycolic acids, like some other genera of aerobic actinomycetes. Akin to other mycolic acid producing aerobic actinomycetes, they are partially acid fast. This staining uses a weaker decolorizer than a traditional acid-fast stain due to the reduced abundance of mycolic acids in the *Nocardia* cell wall. This cell wall feature can be exploited for a rapid genus-level presumptive identification when paired with colony morphology and Gram stain. The presence of mycolic acids in the cell wall occludes crystal violet in the Gram stain and results in the "beaded" appearance mentioned above and seen in Fig. 1.20.

The colonies of *Nocardia* are typically described as dry, white, and powdery with a distinct "earthy" or dirt-like aroma that can be overpowering in the incubator and benchtop. While these features are common with many species of *Nocardia*, there is a great diversity of colony morphology within the genus (and within species), including smooth, irregular (resembling coral), wrinkled, matte, velvety, and rough. Likewise, while white and chalky are the most common color and consistency, other colors can be seen in *Nocardia* colonies including peach and pale yellow. The reverse of the colonies (when viewed through translucent media) can appear brown, tan, yellow, pink, purple, orange, gray, or white. These features are not appreciated on opaque bacteriology media such as blood agar and chocolate agar plates. When growth is visualized only from the front of the plate, the aerial hyphae can obscure the pigment produced by many colonies. Historically, *Nocardia* were considered filamentous molds and their growth was often sought on mycology solid media, which are typically translucent and allowed for visualization from the reverse side of the plate.

The majority of *Nocardia* infections are associated with traumatic inoculation or inhalation (especially in immunocompromised hosts) resulting in pulmonary disease, often with hematogenous spread to other sites including the brain. *Nocardia brasiliensis* specifically has been associated with wound infections, particularly in the setting of traumatic accidents that are contaminated with soil, sand, or other outdoor debris. Though most infections occur in patients with impaired immune response or some degree of immune suppression; otherwise, healthy individuals have also experienced abscesses and cutaneous infections. Typically, immunocompetent hosts can recover much more quickly from these infections, especially with appropriate antibiotic therapy. *N. brasiliensis* is thought to be the most common cause of actinomycotic mycetoma in the New World though cases have been described in many regions of the world. The most common climates associated with infection are those that are hot and dry. In North America, Mexico and the US Southwest are common areas of acquisition. This species has been particularly associated with mycetoma, cellulitis, abscess formation, and lymphocutaneous disease. Many clinical cases have been attributed to traumatic experiences such as falls, accidents, and crashes; however, associations with lesser trauma to the skin have also been documented such as insect bites, cat scratches, and punctures by various plant thorns. As discussed above, with the evolving aspect of *Nocardia* taxonomy, there is evidence that previously ascribed *N. brasiliensis* cutaneous infections may have been caused by *N. pseudobrasiliensis*, with the former being more commonly associated with mycetoma.

Culture remains the most common method of detection and identification of nocardiosis. For wound and tissues cultures, blood cultures, and various sterile sites, routine laboratory agar will support the growth of this genus, including blood and chocolate agar (Fig. 1.21). Media typically used for fungi and acid-fast bacilli will also support the growth of *Nocardia* (e.g., Sabouraud dextrose agar and Löwenstein-Jensen medium). For wound and respiratory specimens, a buffered charcoal yeast extract plate can also be helpful for the recovery of *Nocardia* (Fig. 1.22) though grossly contaminated sources (e.g., poorly cared for wounds, sputum, or tracheal aspirate specimens from cystic fibrosis patients or drowning victims) should be plated on selective buffered charcoal yeast agar containing colistin, cephalothin, cycloheximide, and vancomycin to improve the recovery of these organisms in the presence of more rapid growing organisms (e.g., *Pseudomonas, Staphylococcus, Klebsiella, E. coli,* filamentous fungi).

Nocardia infections, especially those in wounds, are empirically treated with sulfonamides, specifically oral trimethoprim-sulfamethoxazole. Critical or

Fig. 1.22 *Nocardia brasiliensis* colonies on buffered charcoal yeast extract media. Inset shows aerial hyphae and chalky appearance of colonies

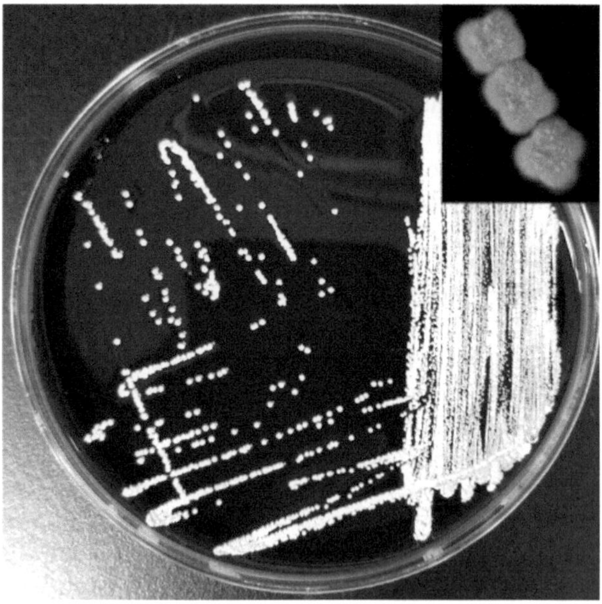

life-threatening infections are also treated with a second drug, typically intravenous, in combination with trimethoprim-sulfamethoxazole. Most isolates of *Nocardia* remain susceptible to trimethoprim-sulfamethoxazole. However, because resistance has been described in a small subset of isolates, antimicrobial susceptibility testing is recommended for all isolates of clinical significance. Broth microdilution is the recommended method for determining susceptibility to antibiotics for *Nocardia*. Clinical Laboratory Standards Institute breakpoints exist for several antimicrobials for *Nocardia*. The primary drugs recommended for testing include amikacin, amoxicillin-clavulanic acid, ceftriaxone, ciprofloxacin, moxifloxacin, clarithromycin, imipenem, linezolid, minocycline, doxycycline, trimethoprim-sulfamethoxazole, and tobramycin.

Summary Points

- *Nocardia* are aerobic actinomycetes that are ubiquitous in the environmental.
- Though technically Gram-positive rods, the Gram stain is often described as beaded and branching due to the presence of mycolic acids preventing the crystal violet from thoroughly penetrating the cell wall.
- *Nocardia* are typically associated with respiratory and wound infections, primarily in immunocompromised hosts and patients on low-dose immunosuppression.
- Culture is the primary method of detection.
- Treatment is typically trimethoprim-sulfamethoxazole with or without a second drug based on clinical indications.

Selected References

Chen BM, Tang J, Lu Z, Wang N, Gao X, Wang F. Primary cutaneous nocardiosis in a patient with nephrotic syndrome. Medicine. 2016;95(3):1–3.

Conville PS, Brown-Elliot BA, Witebsky FG. Chapter 30. *Nocardia, Rhodococcus, Gordonia, Actinomadura, Streptomyces,* and other aerobic actinomycetes. In: Carroll KC, Pfaller MA, editors. Manual of clinical microbiology. 12th ed. Washington, DC: ASM Press; 2019. p. 525–57.

Fukuda H, Saotome A, Usami N, Urushibata O, Mukai H. Lymphocutaneous type of nocardiosis caused by *Nocardia brasiliensis*: a case report and review of primary cutaneous nocardiosis caused by *N. brasiliensis* reported in Japan. J Dermatol. 2008;55(6):346–53.

Smego RA, Gallis HA. The clinical spectrum of *Nocardia brasiliensis* infection in the United States. Rev Inf Dis. 1984;6(2):164–80.

Case 4

Cameron A. Brown and James J. Dunn

A 4-year-old previously healthy male began to have fevers up to 102 °F about 1 week prior to presenting to the hospital. Initially, the fever did respond to antipyretics (e.g., Tylenol) but would return when the child was off medication. During this time, he was more tired than usual but had no specific complaints. His immunizations were up to date, and he had no significant past medical or surgical history. He lived in North Texas with his parents and four siblings and the family had not traveled recently. Nor did he have any exposures to animals or any unusual outdoor environments.

The child was eventually brought to the hospital after waking up with fever of 104 °F, complaining of headache and displaying incoherent speech. He vomited three times en route to the hospital where he was found to have nuchal rigidity and reported pain when his neck was moved. He was confused and disoriented. These findings were consistent with a central nervous system infection so a lumbar puncture was performed to obtain cerebrospinal fluid (CSF). The white blood cell (WBC) count in CSF was 1076 cells/mm^3 (reference range, 0–5 cells/mm^3) with 80% neutrophils, 12% lymphocytes, and 8% monocytes. There were 118 red blood cells (RBC)/mm^3 (reference range, 0 RBC/mm^3) and CSF protein was measured at 125 mg/dL (reference range, 15–45 mg/dL) and glucose at 21 mg/dL (reference range, 40–60 mg/dL).

Empiric parenteral antibiotic treatment with ceftriaxone was initiated along with acyclovir for coverage of potential herpes simplex virus (HSV) central nervous system infection. No organisms were observed on Gram stain of the CSF. A rapid (1 h), highly multiplexed molecular panel for central nervous system pathogen testing was performed. This panel includes a number of common bacterial, viral, and fungal

Table 1.3 Multiplex meningitis/encephalitis panel for rapid, direct testing of CSF specimens

Bacterial targets	Viral targets	Fungal target
Escherichia coli K1	Cytomegalovirus (CMV)	*Cryptococcus neoformans/gattii*
Haemophilus influenzae	Enterovirus	
Listeria monocytogenes	Herpes simplex virus (HSV)-1	
Neisseria meningitidis	Herpes simplex virus (HSV)-2	
Streptococcus agalactiae	Human herpes virus (HHV)-6	
Streptococcus pneumoniae	Parechovirus	
	Varicella zoster virus (VZV)	

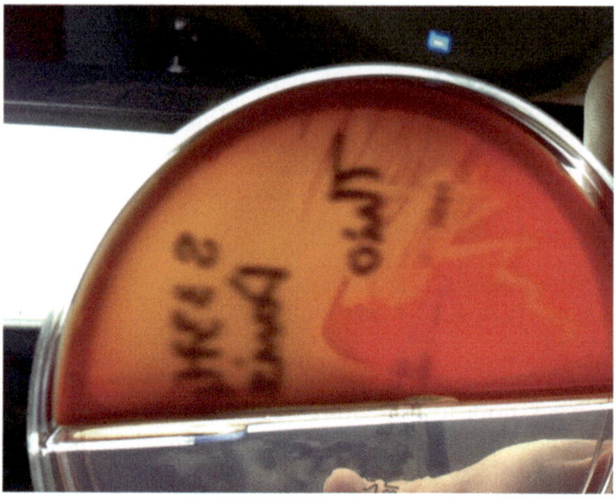

Fig. 1.23 Growth of *L. monocytogenes* on 5% sheep blood agar

pathogens that are known to cause central nervous system infection (Table 1.3). The patient CSF sample was positive using this assay for *Listeria monocytogenes*. Ceftriaxone was therefore discontinued and intravenous ampicillin and gentamicin were started. The following day, the CSF culture plates showed growth with small, smooth colonies that displayed a narrow zone of β-hemolysis (Fig. 1.23). Gram stain of the colonies were small, regular Gram-positive rods (Fig. 1.24) and MALDI-TOF mass spectrometry identified the organism as *L. monocytogenes*.

Discussion

L. monocytogenes is a Gram-positive, motile, facultative anaerobic bacillus. It grows best at 37 °C but can replicate at temperatures from 0 to 50 °C. The organism is broadly distributed in the environment in soil, water, and vegetation. Due to its widespread distribution and the ability to grow at 4 °C, *L. monocytogenes* has been found to be a common contaminant of a variety of both raw and processed foodstuffs. Large foodborne outbreaks have implicated a number of sources including

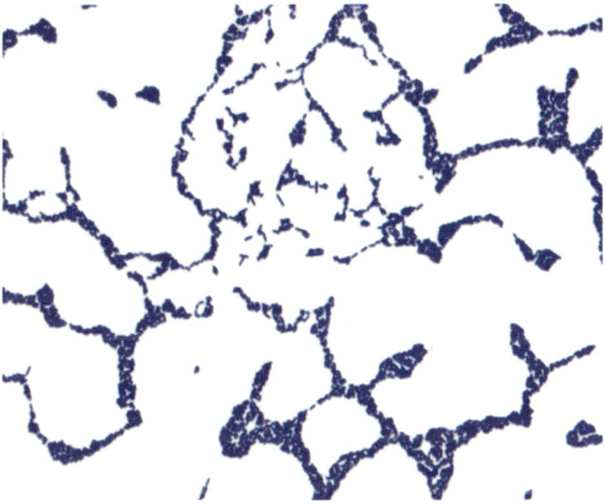

Fig. 1.24 Gram stain from culture plate of *L. monocytogenes* showing short, regular Gram-positive bacilli

dairy products (e.g., cheese, milk, ice cream), fruits and vegetables (e.g., cabbage, lettuce, blueberries), and meat products (e.g., deli meats, uncooked hot dogs, foie gras). Ingestion of the organism by animals or humans may result in clinical disease or transient asymptomatic gastrointestinal colonization. The rate of sporadic human listeriosis in Europe and North America is usually <1/100,000 population per year. In the United States, there are approximately 1600 cases of human listeriosis reported annually with about 250 resulting in death.

A wide variety of clinical syndromes have been associated with *L. monocytogenes* infection. The populations at highest risk for invasive infection include pregnant women, immunocompromised individuals, neonates, and those greater than 60 years of age. Pregnant women infected with *L. monocytogenes* may have a transient bacteremia with flu-like symptoms that is usually self-limiting. However, since *L. monocytogenes* can cross the placenta, it can infect the fetus and cause preterm labor, abortion, stillbirth, or neonatal infection. Infected newborns often have severe disease with bacteremia and/or meningitis. In immunocompromised patients and older adults, *L. monocytogenes* infection can cause sepsis, meningoencephalitis, rhomboencephalitis, hepatitis, liver abscesses, and endocarditis. Cases of *Listeria* gastroenteritis are seen during foodborne outbreaks and symptoms usually manifest 24 h after ingestion of a food item that is contaminated with a large number of bacteria (10^5–10^9 cfu/mL). During these epidemics of foodborne listeriosis, rare cases of sepsis and/or meningitis have been documented in apparently healthy individuals of all ages.

Diagnosis of *L. monocytogenes* infection relies on isolation of the organism in culture or detection using nucleic acid amplification from a normally sterile site. Gram stain of CSF is frequently negative in cases of culture-proven meningitis due to low concentration of organism, so rapid assays for detection of *L. monocytogenes* DNA can be important for patient management. *L. monocytogenes* grows on routine

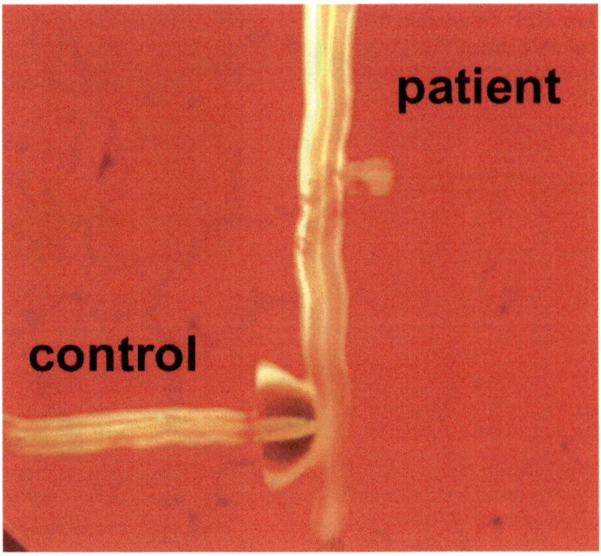

Fig. 1.25 CAMP test on 5% sheep blood agar. The patient isolate shows enhanced hemolysis near the vertical *S. aureus* streak. The control organism (*Streptococcus agalactiae*) also displays enhanced hemolysis

types of media and after 24 h of incubation the colonies are small, smooth, and typically show a narrow zone of β-hemolysis. The organism is catalase positive and hydrolyzes esculin. At 20–28 °C, *L. monocytogenes* displays the characteristic tumbling motility by means of one to six peritrichous flagella. The organism is CAMP test positive. This test involves plating a streak of β-hemolytic *S. aureus* with a perpendicular streak of *L. monocytogenes* up to but not touching the *S. aureus*. Enhanced hemolysis of *L. monocytogenes* near the *S. aureus* streak indicates a positive test (Fig. 1.25).

Treatment of *L. monocytogenes* meningitis involves combination therapy using ampicillin and a second agent such as gentamicin which act synergistically to kill the organism. The duration of treatment is at least 21 days. Alternative antibiotics that also have activity against *L. monocytogenes* include trimethoprim-sulfamethoxazole, linezolid, and rifampin. Importantly, the organism is intrinsically resistant to cephalosporins; the type of antibiotic often used as empiric therapy in cases of suspected bacterial meningitis.

Summary Points

- *Listeria monocytogenes* is a Gram-positive rod that inhabits a broad ecologic niche including soil, water, and vegetation.
- Humans typically become infected with *L. monocytogenes* through ingestion of contaminated food.
- Widespread foodborne outbreaks of *L. monocytogenes* infection can occur due to contamination of dairy, meat, and fruit/vegetable products.

- Immunocompromised individuals, pregnant women, and those over 60 years of age have up to 100- to >1000-fold increased risk of developing listeriosis.
- The most common clinical syndromes associated with *L. monocytogenes* infection are meningitis, bacteremia, and gastroenteritis.
- In cases of *Listeria* meningitis, rapid identification of the organism is critical to ensure administration of appropriate antibiotics.
- *L. monocytogenes* is intrinsically resistant to cephalosporin antibiotics.

Selected References

Arslan F, Meynet E, Sunbul M, et al. The clinical features, diagnosis, treatment, and prognosis of neuroinvasive listeriosis: a multinational study. Eur J Clin Microbiol Infect Dis. 2015;34:1213–21.

Ben Shimol S, Einhorn M, Greenberg D. Listeria meningitis and ventriculitis in an immunocompetent child: case report and literature review. Infection. 2012;40:207–11.

Schlech WF. Epidemiology and clinical manifestations of Listeria monocytogenes infection. Microbiol Spectr. 2019;7(3). https://doi.org/10.1128/microbiolspec.GPP3-0014-2018.

Case 5

Angela Ma and Marc Roger Couturier

A 35-year-old male presented to his primary care physician with a several month history of vague, nondescript epigastric pain. The man had no family history of ulcers or gastric cancer, and no immediate alarm symptoms as defined by the American College of Gastroenterology (e.g., unexplained weight loss, jaundice, gastric bleed). The physician performed an abdominal ultrasound and ordered a complete blood count (CBC), chemistry panel, and noninvasive stool antigen test (SAT) for *Helicobacter pylori*. No abnormalities were observed with the abdominal ultrasound and the results from the CBC and chemistry panel were within normal ranges. However, the SAT for *H. pylori* returned positive, and the patient was thus treated empirically with first-line triple therapy consisting of a proton pump inhibitor, clarithromycin, and amoxicillin. One and a half months after completion of therapy, the symptoms had not resolved and the patient was now experiencing intermittent nausea as well. The SAT was repeated and was still positive. As a result of this concern for treatment failure, he was referred to a gastroenterologist. The patient had no documented penicillin allergy, recent history of drugs or alcohol use, or antimicrobial therapy. An esophagogastroduodenoscopy was performed where

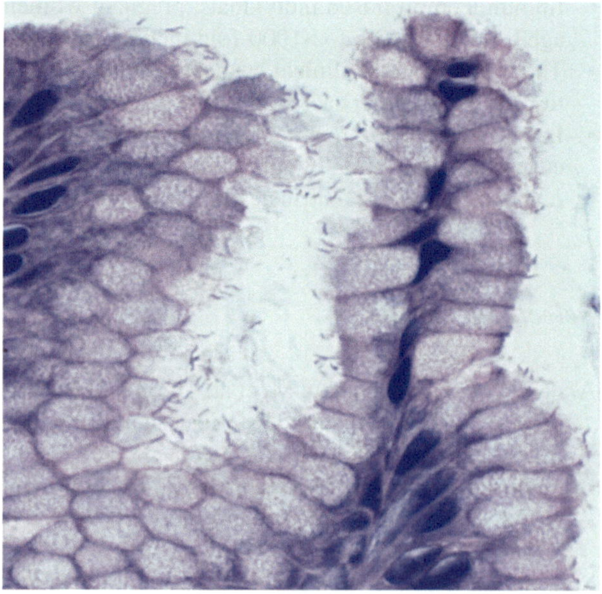

Fig. 1.26 Hematoxylin and eosin-stained gastric antral mucosa revealing numerous curved bacilli, representing *H. pylori* (1000× magnification). (Image courtesy of Gillian Hale, University of Utah Department of Pathology)

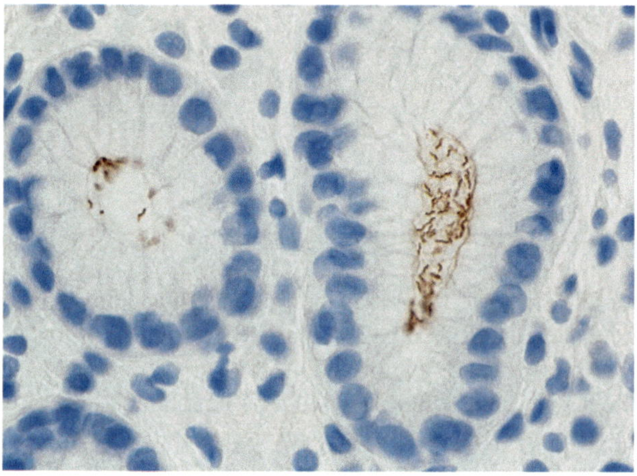

Fig. 1.27 Immunohistochemical stain for *H. pylori* gastric mucosa revealing numerous curved *H. pylori* within a crypt (1000× magnification). (Image courtesy of Gillian Hale, University of Utah Department of Pathology)

antral gastritis was observed and multiple biopsies were obtained for histologic workup and microbiological testing. Anatomic pathology reported curved bacilli resembling *H. pylori* with neutrophilic infiltrates that were observed in multiple tissue sections stained with hematoxylin and eosin, suggestive of acute infection (Fig. 1.26). An *H. pylori* immunohistochemical stain was also positive for numerous organisms resembling *H. pylori* (Fig. 1.27). Two gastric biopsy specimens obtained during the endoscopy were preserved in enteric transport media and sent to a

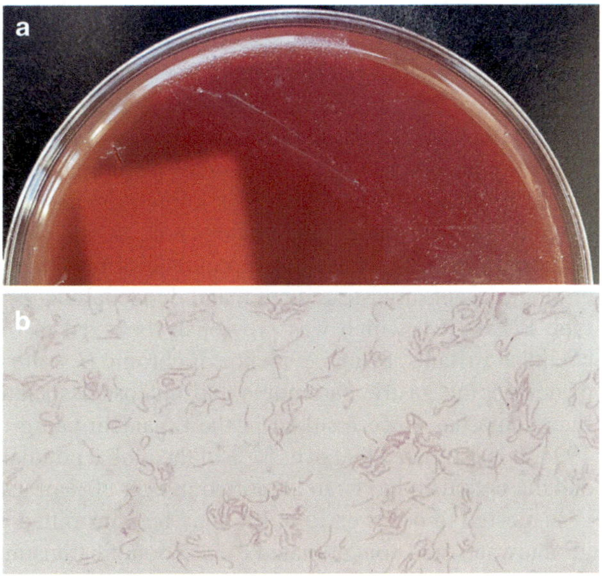

Fig. 1.28 Pure culture of *H. pylori* on sheep blood agar (**a**) and (**b**) Gram stain demonstrating Gram-negative curved bacilli captured at 1000× magnification

reference laboratory for microbiology culture and antimicrobial susceptibility testing. Specimens were homogenized in 1 mL of tryptic soy broth and then plated onto Brucella agar plates. Agar plates were incubated in a microaerobic (5% O_2, 10% CO_2, 85% N) environment at 35 °C and examined for growth after 3-, 5-, and 7-days incubation. At day 5, small, translucent colonies were observed on solid media and Gram stains correlated with *H. pylori* morphology (Fig. 1.28). MALDI-TOF mass spectrometry confirmed the identity of the organism as *H. pylori*.

Antimicrobial susceptibility testing was performed using agar dilution, where it was determined that the isolate was resistant to clarithromycin (minimum inhibitory concentration of 4.0 μg/mL) but susceptible to amoxicillin, levofloxacin, metronidazole, tetracycline, and rifampin. After obtaining antimicrobial susceptibility testing results, the patient's gastroenterologist decided to treat the patient with a 14-day course of bismuth quadruple therapy with omeprazole, bismuth subsalicylate, tetracycline, and metronidazole. Patient compliance with therapy was followed closely by his care provider and symptoms of gastritis improved over the course of treatment. Eradication of *H. pylori* was confirmed with follow-up testing by gastroenterology using urea breath test 4 weeks after completion of therapy, which was negative.

Discussion

Identified in 1984, *H. pylori* is a motile, curved bacilli possessing 3–7 lophotrichous flagella that had previously been classified as a member of the *Campylobacter* genus. Similar to *Campylobacter* spp., *H. pylori* is a Gram-negative, microaerophilic rod-shaped organism. The morphology of *H. pylori* often changes to a coccoid shape following the use of proton pump inhibitors, eradication therapy, and adverse environmental conditions. This morphology can be appreciated in aging cultures as well as direct histologic examination. *H. pylori* is fastidious and while it does not require glucose for growth, its survival is significantly enhanced in the presence of various carbon sources including glucose, pyruvate, succinate, and citrate. Infection with *H. pylori* is known to cause a chronic superficial gastritis, marked by a low-grade inflammatory process, but commonly presents as an asymptomatic infection or mildly symptomatic in upwards of 90% of patients. It is the etiologic agent of gastric, peptic, and duodenal ulcers and strongly associated with type B gastritis, which was previously believed to be associated with stress and chemical irritants. Strong evidence of chronic *H. pylori* infection and increased risk of developing gastric carcinoma and mucosa-associated lymphoid tissue (MALT) lymphoma has since resulted in the organism being classified as a carcinogen in 1994. It is estimated that over 50% of the global population is infected with *H. pylori* and the organism has been identified in over 80% of gastric ulcer patients. Infection is suggested to occur early in life may be transmitted by oral-oral, fecal-oral, fresh groundwater, and contaminated food. Zoonotic transmission of *H. pylori* is also currently under debate as the organism has been isolated from domestic and ruminant animals. *H. pylori* colonizes the antrum of the stomach and adheres to the gastric mucosa for replication. The organism is capable of disrupting tight junctions between gastric epithelial cells to access the basement membrane and has been demonstrated in *in vitro* and *in vivo* studies to invade gastric mucosa lesions. Through the production of the urease enzyme, *H. pylori* is able to ensure its survival in the low pH conditions of the stomach. Outside of the stomach, growth is limited for *H. pylori* and instead, the organism typically enters a viable but non-culturable state. While other species of *Helicobacter* including *H. heilmannii*, *H. cinaedi*, and *H. fennelliae* differ from *H. pylori* in the lack of correlation with gastric carcinoma, they are causative agents of gastroenteritis.

Standard first-line therapy for *H. pylori* infection consists of a proton pump inhibitor, clarithromycin, and amoxicillin or metronidazole for 14 days in patients with no previous antibiotic exposure. Clarithromycin resistance is the most common cause of treatment failure with first-line therapy and is observed in 17.4% of cases in the United States. Patients with previous macrolide exposure, penicillin allergies, or clarithromycin-resistant *H. pylori* may require bismuth quadruple therapy, which consists of a proton pump inhibitor, bismuth, tetracycline, and a nitroimidazole. Other therapy regimens can be recommended based on the patient's adherence to therapy, tolerance to side effects, and antimicrobial susceptibility profiles of *H. pylori*.

A number of non-culture-based, noninvasive methods are available and preferred for the diagnosis of non-complicated, active *H. pylori* infection. Stool antigen enzyme immunoassays using monoclonal or polyclonal capture antibodies against *H. pylori* antigens in either a rapid immunochromatographic assay or 96 well ELISA have high sensitivity and specificity for diagnosis and test-of-cure. Tests using pooled monoclonal antibodies have shown the best performance characteristics of the stool antigen tests. The urea breath test (UBT) is an alternative to the SAT for active *H. pylori* infection. The UBT is a noninvasive method that exploits the organism's ability to convert urea to CO_2 and NH_3 and is recommended for treatment-naive patients with no active peptic ulcer bleeding. The test can be used for both diagnosis and test-of-cure. Patients breathe into a collection bag to collect a pre-test baseline specimen followed by swallowing a liquid containing urea labeled with nonradioactive carbon-13. The urea is subsequently degraded through the urease activity of *H. pylori* in the stomach, resulting in isotope-labeled CO_2 and NH_3. The labeled 13-CO_2 isotope is absorbed into the blood stream and thus detectable in exhaled breath when measured by an infrared spectrophotometer. Patients with *H. pylori* infection will have an increased ratio of 13-CO_2 to the standard carbon isotope 12-CO_2 in the control specimen. The sensitivity and specificity of the UBT is reported to range from 88–95% and 95–100%, respectively.

In circumstances of active peptic ulcer bleeding or a need for biopsy following recent therapy, gastric tissue is the recommended specimen type. Histologic examination of gastric biopsies stained with hematoxylin and eosin can be sufficient to detect the presence of the organism for diagnosis of *H. pylori* infection. Additional special stains and immunohistochemistry can also be used but are recommended in circumstances where the organism was not detected by hematoxylin and eosin staining and should not be performed as part of the initial histologic workup. Histologic stains do not differentiate *H. pylori* from other *Helicobacter*- or *Campylobacter*-like species and commercial antibodies used for *H. pylori* immunohistochemistry often cross-react *H. heilmannii* as well (Fig. 1.29). A commercial rapid urease test can also be performed on biopsy specimens and the test performs well when high numbers of *H. pylori* are present in the tissue. The biopsy specimen is placed in a

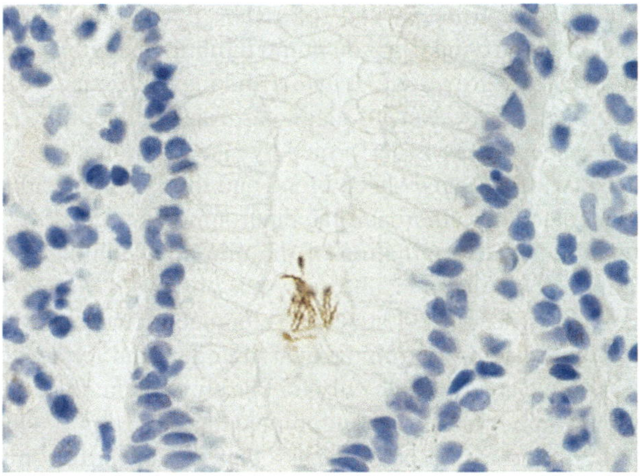

Fig. 1.29 Immunohistochemical stain for *H. pylori* showing cross-reactive staining of *H. heilmannii* in the gastric antral mucosa (1000× magnification). Note that *H. heilmannii* is longer than *H. pylori* and tightly spiraled. (Image courtesy of Gillian Hale, University of Utah Department of Pathology)

substrate containing urea and phenol red as a pH indictor that will change color from yellow to red due to production of ammonia following reaction of urea with urease produced by *H. pylori*. This assay, similar to the UBT, will not distinguish *H. pylori* from other urease-positive gastric *Helicobacter* spp.

Serologic assays are available to detect IgG antibodies produced against *H. pylori* but they offer little clinical value in diagnosing an active infection or to assess whether *H. pylori* has been eradicated following therapy as immunoglobulins can remain elevated in patients who have successfully cleared the infection. As a result, this testing modality is no longer recommended in most guidelines for the majority of patients.

Culture of *H. pylori* is very specific but sensitivity can be poor compared to other methods due to the fragile nature of the organism. Proper specimen collection and transport prior to culturing is critical to recovery of *H. pylori*; duodenal or gastric biopsies should be preserved in sterile saline, Stuart's medium, brucella broth with glycerol, or enteric transport media such as Cary-Blair and delivered to the microbiology laboratory as soon as possible without large fluctuations in atmospheric or temperature conditions to minimize the loss of viable cells. Specimens are typically plated onto enriched, nonselective agar to facilitate growth of *H. pylori* but in the case of contaminated gastric specimens, antibiotic-supplemented agar such as Skirrow's or Dent's agar may be required to inhibit the growth of other microflora. Agar plates are incubated at 35–37 °C in microaerobic conditions of 5–10% O_2 and 5–15% CO_2 for a period of 3–7 days. *H. pylori* colonies have a small, translucent, and circular appearance on agar and typically take at least 48 h to grow. Common biochemical properties of *H. pylori* that can be tested for in the laboratory include catalase (positive), urease (positive), and nitrate reduction (negative) activity, but they are not distinctive to sufficiently aid in identification of *H. pylori* from other organisms. More commonly used methods for identification of *H. pylori* from culture are MALDI-TOF spectrometry and 16S rRNA gene sequencing. It is important to note that use of antimicrobials, proton pump inhibitors, and bismuth preparations are known to suppress the growth of *H. pylori* and may subsequently result in false-negative test results for all methods of diagnosis, especially culture.

Antimicrobial susceptibility is performed by agar dilution using Mueller-Hinton agar supplemented with 5% aged sheep blood and thus requires cultured *H. pylori*. The plates are incubated in a microaerobic environment at 35–37 °C and read after 3 days of incubation for determination of the minimum inhibitory concentration, which is the dilution at which no visible growth is observed. *H. pylori* antimicrobial susceptibility testing is performed at select reference laboratories for amoxicillin, clarithromycin, levofloxacin, metronidazole, rifampin, and tetracycline. However, interpretive breakpoints are available only for clarithromycin. In addition to culture-based antimicrobial susceptibility testing, molecular detection of single nucleotide variations in the 23S ribosomal RNA gene associated with clarithromycin resistance can also be performed on stool specimens and gastric biopsies from patients with *H. pylori* infection.

Summary Points

- *H. pylori* is a class 1 (definite) carcinogen associated with increased risk for development of gastric cancer and MALT lymphoma.
- Patients can be diagnosed with *H. pylori* infection by non-culture-based methods including urea breath test, stool antigen test, and histologic staining and rapid urease tests of biopsy specimens collected during endoscopy.
- Standard therapy regimens consist of a proton pump inhibitor, clarithromycin, and amoxicillin or metronidazole. In cases of clarithromycin resistance, penicillin allergies, or treatment failure, alternative therapy options are used.
- Clarithromycin resistance is common, has been observed globally, and has increased to 17.4% in the United States.

Selected References

Batts KP, Ketover S, Kakar S, Krasinskas AM, Mitchell KA, Wilcox R, Westerhoff M, Rank J, Gibson J, Mattia AR, Cummings OW, Davison JM, Naini BV, Dry SM, Yantiss RK; Rodger C Haggitt Gastrointestinal Pathology Society. Appropriate use of special stains for identifying *Helicobacter pylori*: recommendations from the Rodger C. Haggitt Gastrointestinal Pathology Society. Am J Surg Pathol. 2013;37(11):e12–22.

Chey WD, Leontiadis GI, Howden CW, Moss SF. ACG clinical guideline: treatment of *Helicobacter pylori* infection. Am J Gastroenterol. 2017;112(2):212–39.

Hooi JKY, Lai WY, Ng WK, Suen MMY, Underwood FE, Tanyingoh D, Malfertheiner P, Graham DY, Wong VWS, Wu JCY, Chan FKL, Sung JJY, Kaplan GG, Ng SC. Global prevalence of *Helicobacter pylori Infection*: systematic review and meta-analysis. Gastroenterology. 2017;153(2):420–9.

Wang F, Meng W, Wang B, Qiao L. *Helicobacter pylori*-induced gastric inflammation and gastric cancer. Cancer Lett. 2014;345(2):196–202. https://doi.org/10.1016/j.canlet.2013.08.016.

Case 6

Cameron A. Brown and James J. Dunn

A 1-year-old male who recently emigrated from Pakistan presented to the emergency department with fevers, persistent unilateral cervical and axillary lymphadenopathy, and progressively worsening cough. His past medical history was significant for recurrent pneumonia, including *Staphylococcus aureus* and *Aspergillus fumigatus*. He had been following the vaccination schedule of Pakistan until he moved to the United States 3 months prior, which included a single bacillus Calmette-Guérin (BCG) vaccine at birth. The patient lives at home with his parents and no other siblings and no pets. The patient was admitted to the hospital for additional evaluation.

Upon admission, the child was found to be febrile at 38.1 °C, oxygen saturation was at 95% on room air and respiratory rate was 45 breaths/min. A chest radiograph was obtained, which showed left lung pleural effusion and consolidation and right lung perihilar opacities (Fig. 1.30). A complete blood count was ordered that showed a leukocytosis of 18 × 10^3 WBCs/μL (Ref: 6.0–17.5 × 10^3 WBCs/μL) with 26.2% neutrophils (ref: 20–48%), 62.5% lymphocytes (ref: 34–88%), and 8.0% monocytes (ref: 0–5%). Given the radiographic findings and leukocytosis, the differential diagnosis included common causes of community-acquired bacterial pneumonia, namely, *Streptococcus pneumoniae* but also *Haemophilus influenzae*, though the patient had received three doses of the PCV10 and Hib vaccines. Given his frequent admissions for pneumonia, causes of hospital-acquired pneumonia such as *Staphylococcus aureus, Pseudomonas aeruginosa,* and Enterobacterales such as *Klebsiella pneumonia* and *Escherichia coli* were also considered. The differential also included *Mycobacterium tuberculosis* (MTB), considering his potential for exposure while in Pakistan, a country with a high incidence of tuberculosis, as well as general lymphadenopathy. Interferon-γ release assays (IGRAs) were repeatedly indeterminate, failing to rule out *M. tuberculosis* exposure. Fungal pneumonia was considered given his previous history, but negative galactomannan results suggested the likelihood of *Aspergillus* infection was low.

Routine bacterial, fungal, and mycobacterial cultures were performed from blood, bronchoalveolar lavage (BAL) fluid, and an excisional biopsy of the left axillary lymph node. Gram stain of the lymph node and BAL fluid revealed no organisms, and routine bacterial and fungal cultures for all three specimens yielded only normal flora from the BAL. Auramine O staining of the BAL fluid after concentration showed numerous bacilli (Fig. 1.31), and Kinyoun stain was positive for

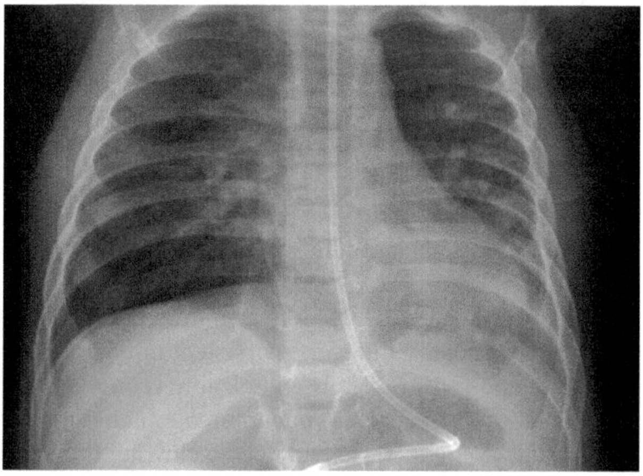

Fig. 1.30 Frontal chest radiograph showing left lung pleural effusion and development of right pleural thickening, in addition to patchy perihilar opacities in the right lung and left lower lobe opacities

Fig. 1.31 Auramine O stained bronchoalveolar lavage showing fluorescent bacilli

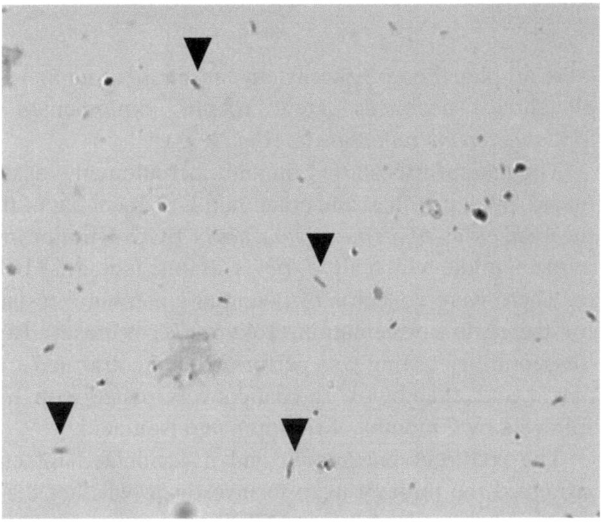

Fig. 1.32 Kinyoun stained bronchoalveolar lavage showing acid-fast bacilli (arrows)

acid-fast bacilli (AFB) (Fig. 1.32). BAL fluid and the lymph node biopsy were both sent to a reference lab for MTB nucleic acid detection, and both were positive for *M. tuberculosis* complex and negative for mutations associated with rifampin resistance. Blood collected in a lysis-centrifugation tube was inoculated into a Middlebrook 7H9 liquid media and placed on an automated instrument to detect growth. The automated instrument indicated growth after 8 days, which also revealed AFB. The BAL fluid, lymph node biopsy, and the positive liquid media

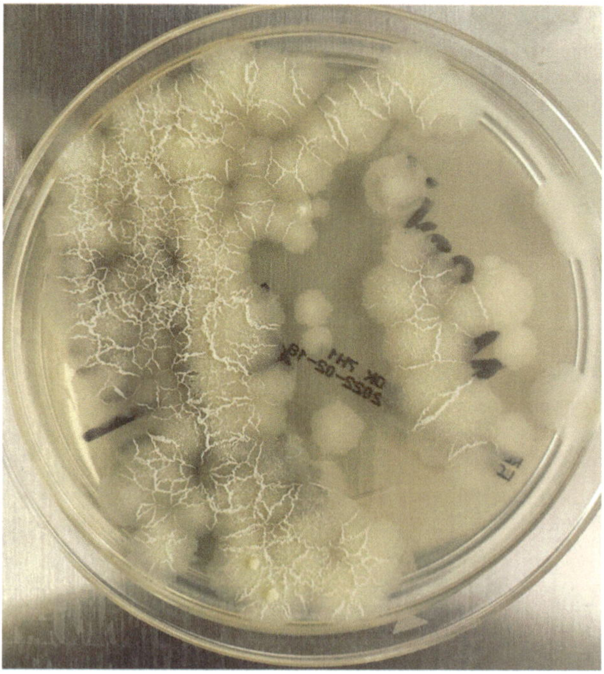

Fig. 1.33 Middlebrook 7H11 showing rough, wrinkled colonies after 3 weeks of incubation at 37 °C

were all plated onto Löwenstein-Jensen medium and Middlebrook 7H11 agar, and all three specimens grew rough, unpigmented colonies consistent with *Mycobacterium tuberculosis* (Fig. 1.33).

The patient was started on antituberculous therapy, consisting of rifampin, isoniazid, pyrazinamide, and ethambutol. Sequencing of the 16S rDNA gene identified the species as *Mycobacterium bovis* BCG. This prompted the discontinuation of pyrazinamide, which all *M. bovis* strains, including BCG, are intrinsically resistant to. There were concerns of potential ethambutol resistance though the strain used for the child's vaccination, Tokyo 172, is generally sensitive to that antibiotic. Susceptibility testing was performed and confirmed susceptibility to rifampin, isoniazid, and ethambutol. The patient was treated with these three drugs for 2 months, followed by 7 months of rifampin and isoniazid.

The recurrent pneumonia and disseminated infection from the BCG vaccine prompted the primary team to investigate whether the patient had an undiagnosed immunodeficiency. A neutrophil function test was performed, and the patient was found to have X-linked chronic granulomatous disease. The patient was started on trimethoprim-sulfamethoxazole and itraconazole for prophylaxis, as well as subcutaneous interferon-γ, while the patient awaited hematopoietic stem cell transplantation.

Discussion

Bacillus Calmette-Guérin (BCG) is a live, attenuated strain of *Mycobacterium bovis* that is used for immunization against tuberculosis (TB), particularly in areas of high prevalence of disease. The original BCG strain was created after *M. bovis* recovered from a cow with mastitis was serially subcultured onto potato slices over the course of many years to create a strain that did not produce disease in various animal models. This strain was then propagated by the Insitut Pasteur and donated to laboratories to produce their own BCG strain for immunization against tuberculosis. Today, there are multiple genetically distinct strains of BCG that are used for vaccinations, each with varying phenotypic properties and immunogenicity. Vaccination with BCG has been shown to decrease the risk of progressing to active disease though protection is thought to wane as the child reaches adolescence.

M. bovis, including the BCG strain, are within the larger *Mycobacterium tuberculosis* (MTB) complex comprised of all the species of *Mycobacterium* that are known to cause human or zoonotic tuberculosis. All species within this complex are thought to share a single common ancestor and have nearly identical genomes. In addition, there are more than 170 other species of *Mycobacterium* other than MTB complex and *M. leprae*, the causative agent of Hansen's disease, that are collectively referred to as non-tuberculous mycobacteria (NTM).

All *Mycobacterium* species have a cell wall that contains long chains of mycolic acid and other complex lipids that prevent the mycobacteria from staining readily with Gram stain, instead giving a "beaded" appearance, where the bacilli stains irregularly, or as a "ghost," where the bacilli does not stain at all but can appear as a clear zone in a stained background. The preferred stains for *Mycobacterium,* with the exception of *M. leprae*, are Ziehl-Neelsen or the Kinyoun stain, both of which use carbolfuchsin as the primary stain and an acid-alcohol to decolorize other organisms. Because *Mycobacterium* retain the carbolfuchsin after washing with acid-alcohol, they are considered "acid-fast," which leads to the synonym of acid-fast bacilli, or AFB, for mycobacteria. A modification of Ziehl-Neelsen or Kinyoun, wherein a weaker acid-alcohol is used to decolorize, can be used to stain other weakly acid-fast organisms, such as *Nocardia* or *Streptomyces*, as well as *M. leprae*. In addition to this stain, a fluorescent stain, auramine O with or without rhodamine B, is frequently used to identify mycobacteria from patient specimens. However, unlike the Kinyoun or Ziehl-Neelsen without modification, *Nocardia* and other weakly acid-fast organisms will fluoresce when stained with auramine O.

The majority of TB infections are acquired from inhalation of respiratory droplets secreted by persons with active, pulmonary TB. The most common etiological agent within the MTB complex is *Mycobacterium tuberculosis,* while *M. bovis* accounts for 1–2% of all cases in the United States and other developed countries. During the primary infection after inhalation of these droplets, if the MTB infection is not immediately eliminated by the innate or acquired immune system, the organism most often enters a latent state inside the host alveolar macrophages and invades the lung interstitial tissue. Other macrophages, dendritic cells, and lymphocytes are recruited to this site and form a granuloma to contain the infection. Most individuals infected with TB will remain in this state, referred to as latent TB infection (LTBI) for decades or more, and will be asymptomatic and noninfectious. Rarely, but usually in young children or HIV-positive patients, MTB infection can immediately progress to active disease without first entering this latent stage.

A small percentage of individuals will progress from LTBI to active disease when the dormant bacteria in the granuloma become reactivated. Impairment of cellular immunity, such as from HIV infection, TNF inhibitors, corticosteroids, and diabetes, have been noted to increase the risk of reactivation, while in highly endemic regions, reinfection with a new strain of MTB can lead to reactivation. Mycobacteria can spread from the previously encapsulated granuloma to infect new areas of the lung within the infected individual or spread to others via respiratory droplets. Patients with active, pulmonary TB may be symptomatic, commonly exhibiting persistent cough, fever, night sweats, and weight loss that have a gradual onset over weeks to months. Radiographic findings indicative of active TB includes nodular infiltrates in the upper lobes of the lung, as well as formation of cavitary lesions.

In extremely rare cases, children who are vaccinated with BCG can develop disease from the vaccine itself, ranging in severity from a localized abscess to disseminated disease. Disseminated BCG, as with all disseminated MTB infections, can involve many organs, including bones, lymph nodes, lungs, liver, and brain as a result of lymphohematogenous spread. Most of these cases are fatal, particularly when the dissemination pattern is miliary. Disseminated BCG and other severe complications most often occurs in children with primary and acquired immunodeficiencies, such as those with severe combined immunodeficiency, chronic granulomatous disease, or HIV. Due to the risk of severe complications, children known or suspected to be immunocompromised are recommended not to receive the BCG vaccine.

Diagnosis of TB exposure can be made with the tuberculin skin test (TST) or an interferon-γ release assay (IGRA). In patients with LTBI or active disease, both the TST and IGRA are positive, but distinguishing between the two disease states is not possible with these tests, and negative results for one or both tests does not rule out TB. Exposed individuals that immediately eliminated MTB may or may not be positive by TST and IGRA. Previous recipients of the BCG vaccine may be falsely positive to the TST but will be negative by IGRA since the latter was specifically designed to exclude BCG strains of *M. bovis*. IGRAs from children less than 2 years old have poor sensitivity and in immunocompromised patients are frequently indeterminate or invalid.

Active pulmonary disease is typically diagnosed with a positive sputum smear or a nucleic acid amplification test (NAAT), followed by AFB culture of the sputum. Given the issues of sensitivity for TB diagnosis, persons suspected of TB should submit three consecutive sputum samples every 8 h for smear and culture. Non-pulmonary TB is more difficult for laboratory diagnosis, depending on the presentation, but a variety of specimen types can be submitted for culture. MTB complex can take 2 or more weeks to form colonies on solid media, such as Löwenstein-Jensen or Middlebrook 7H11. Growth in liquid media can be variable, depending on the initial inoculum of AFB in the sample. *Mycobacterium tuberculosis* are typically described as rough, buff, and unpigmented. In contrast, *M. bovis* is smooth in appearance, while most strains of BCG have a rough appearance. In addition to NAATs that can be performed from direct patient specimens, MALDI-TOF and high-pressure liquid chromatography have all been used to identify individual species of MTB complex and NTM. Strain typing of MTB complex can be performed at the U.S. Centers for Disease Control and Prevention.

Treatment of TB disease consists of a multidrug regimen for 6–9 months. Active TB disease is commonly treated with a combination of rifampin, isoniazid, pyrazinamide, and ethambutol. *M. bovis*, however, is intrinsically resistant to pyrazinamide, and some strains of BCG are also resistant to isoniazid. Furthermore, many strains of *Mycobacterium tuberculosis* are multidrug-resistant, having resistance to rifampin and isoniazid, and require the use of second-line antibiotics. Because of this, it is important to perform susceptibility testing on MTB isolates to help guide therapy.

Summary Points

- *Mycobacterium bovis* is an acid-fast organism that is part of the *Mycobacterium tuberculosis* complex. It can cause infection in humans and animals, resulting in latent tuberculosis or active tuberculosis. Only individuals with active disease are considered infectious.
- Bacillus Calmette-Guérin (BCG) is a live, attenuated strain of *M. bovis* that is given to newborns in many countries to protect from active tuberculosis. Because it is a live organism, newborns with immunodeficiencies can become infected with BCG, potentially resulting in severe complications such as disseminated disease.
- Tuberculin skin test or an interferon-γ release assay can be used to screen individuals for latent or active tuberculosis disease, but sputum smear or culture should be used to diagnose active disease. Culture can take many weeks but is the most sensitive method for detecting live *Mycobacterium* from patient samples.
- Treatment consists of a multidrug regimen that should be taken for 6–9 months to clear the infection. However, given the increasing rates of multidrug-resistant *M. tuberculosis*, susceptibility testing against first-line antituberculous agents should be obtained.

Selected References

Hesseling AC, Rabie H, Marais BJ, et al. Bacille Calmette-Guérin vaccine-induced disease in HIV-infected and HIV-uninfected children. Clin Infect Dis. 2006;42(4):548–58.

Nunes-Alves C, Booty M, Carpenter S, et al. In search of a new paradigm for protective immunity to TB. Nat Rev Microbiol. 2014;12:289–99.
Pai M, Behr M, Dowdy D, et al. Tuberculosis. Nat Rev Dis Primers. 2016;2:16076.
Ritz N, Tebruegge M, Connell TG, et al. Susceptibility of mycobacterium bovis BCG vaccine strains to antituberculous antibiotics. Antimicrob Agents Chemother. 2009;53(1):316–8.

Case 7

Marc Roger Couturier

A previously healthy 15-year-old male presented to the emergency department at the local children's hospital with a 2-day history of 40 °C fever. The patient also had a large 5-inch wound on his calf that had been completely sutured but was extremely painful and "hot" to the touch. During the previous weekend (4 days before admission), the patient sustained a traumatic injury to his calf while recreating in a lake in northern Utah. The patient had attempted to untangle a waterskiing tow rope from the propellor of his boat, when the throttle was accidentally engaged, resulting in the propellor cutting his calf. The patient's family wrapped the wound with towels and mostly stopped the bleeding. Eight hours later, the wound was still bleeding intermittently and had a significant opening, prompting the patient to seek medical care at a rural urgent care clinic. The wound was cleaned, multiple sutures were applied, and the foot was immobilized in a split to prevent the wound from further opening. The patient was prescribed amoxicillin and instructed to follow up with their primary care physician after the weekend when they returned home. At the children's hospital, opaque exudate with blood was seen at the wound site. The immediate tissue and skin surrounding the wound showed dark red and purple discoloration. This appearance radiated from the sutures, extending down toward the foot and upwards toward the knee.

The sutures were removed from the wound at the emergency department and several sections of now necrotic tissue were removed. Tissue specimens from the healthy margin were also collected and both the necrotic and healthy tissues were sent for bacterial tissue culture. Pus from the wound was also submitted for wound cultures. Amoxicillin was discontinued in favor of vancomycin and piperacillin-tazobactam. All specimens were also cultured for anaerobes out of concerns for *Clostridium perfringens.* After 24 h of incubation, large gray beta hemolytic colonies were seen for all cultures (aerobic and anaerobic from both the pus and tissue) on blood agar plates (BAP) (Fig. 1.34a), MacConkey agar (MAC) (Fig. 1.34b), and chocolate agar plates, but no growth was seen on the colistin nalidixic acid (CNA) plates. Gram stain revealed Gram-negative rods with rounded ends approximately 1–2 mm in length, some in short chains or pairs. The colonies were positive for catalase, oxidase, and indole. Colonies were referred to MALDI-TOF mass spectrometry for identification. The identification from the necrotic tissue yielded *Aeromonas caviae* complex and the healthy tissue revealed *Aeromonas hydrophila* complex.

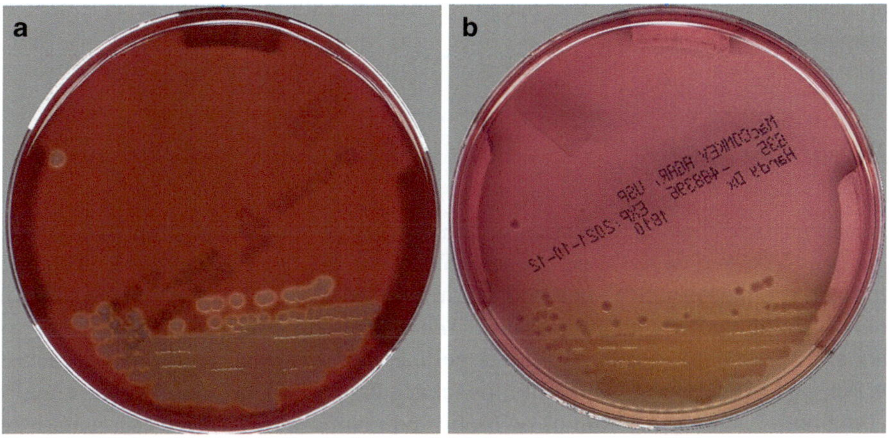

Fig. 1.34 Primary culture plates from the tissue culture. Panel (**a**) shows growth on Columbia blood agar plate with beta-hemolysis. Panel (**b**) shows growth on MacConkey agar plate

Table 1.4 Antimicrobial susceptibility testing profile for two isolates of *Aeromonas* obtained from the patient's leg wound

	Aeromonas caviae complex		*Aeromonas hydrophila* complex	
Antimicrobial	Minimum Inhibitory Concentration (mg/mL)	Interpretation[a]	Minimum Inhibitory Concentration (mg/mL)	Interpretation[a]
Piperacillin-tazobactam	32/4	I	>128/4	R
Cefepime	1	S	1	S
Cefoxitin	>32	R	>32	R
Ceftriaxone	0.5	S	2	I
Meropenem	0.5	S	0.5	S
Gentamicin	1	S	2	S
Aztreonam	2	S	1	S
Ciprofloxacin	1	S	1	S
Trimethoprim-sulfamethoxazole	1/19	S	2/38	S

[a] Breakpoints applied based on the CLSI M45 3rd edition

Isolates were retested and the same identifications were obtained. Isolates were submitted for antimicrobial susceptibility testing and the following susceptibility profiles were obtained (Table 1.4).

The patient was further managed with multiple irrigations and surgical debridement along with meropenem administered intravenously for 10 days. Unfortunately, due to delays in appropriate care and empiric therapy, the patient developed necrotizing fasciitis that was too advanced to salvage the lower leg. A below-the-knee amputation was required. The patient recovered without further debridement or amputations.

Discussion

Aeromonas spp. belong to the family *Aeromonadaceae* which is comprised of three genera. Only *Aeromonas* is pathogenic to humans. The genus *Aeromonas* is characterized as Gram-negative short rods with rounded ends that can occur singular, in pairs, and sometimes in short chains. *Aeromonas* can inhabit a wide diversity of aquatic ecosystems including clean or polluted lakes, swamps, and rivers, brackish water, and even marine water with low concentrations of salt. *Aeromonas* are also common inhabitants of biofilms on biotic and abiotic surfaces such as aquatic vegetation, deadfall/driftwood, docks, boats, and moorings. *Aeromonas* is generally considered an environmental bacterium and is considered an incidental pathogen in humans. The three most well-defined clinical associations with *Aeromonas* infections are wound infections including necrotizing fasciitis, gastrointestinal infections/diarrhea, and soft tissue infections or bacteremia post-leech therapy.

Gastrointestinal infections are typically attributed to ingestion of contaminated food or water though the clinical significance of *Aeromonas* with gastrointestinal infection is not always clear. While *Aeromonas* can be recovered in stool culture in the absence of other pathogens, it can also be recovered in healthy individuals. Patients with gastrointestinal infections can experience mild to severe illness, including dysentery, hemolytic uremic syndrome, and cholera-like illness. Approximately 15–20% of patients with dysentery or cholera-like illness are coinfected with other invasive gastrointestinal pathogens (mainly *Vibrio*), making these clinical associations difficult to exclusively ascribe to *Aeromonas*.

Leech therapy has been used for wound therapy for centuries. Unfortunately, *Aeromonas* is a common inhabitant of leech guts, and the organism can unintentionally be introduced into a surgical wound because of this modern application of therapy. This typically results in bacteremia and/or infection of the surgical wound. Patients will often receive prophylactic antibiotics prior to leech therapy and the leech colonies are screened by culture for any multidrug-resistant *Aeromonas* spp. prior to use.

Wound infections with *Aeromonas* are typically the result of a penetrating wound with an abiotic object that is colonized by an *Aeromonas*-containing biofilm. Open wounds that are exposed to *Aeromonas*-inhabited water can also result in opportunistic infection though the former is a significantly higher risk for infection since high concentrations of bacteria are directly introduced into a wound. These wounds must be thoroughly cleaned, and it is recommended that patients receive appropriate prophylactic therapy to cover *Aeromonas* and other opportunistic pathogens that reside in fresh water. These management decisions depend on the nature of the wound and the ability to thoroughly clean it promptly. Serious complications of untreated or inappropriately treated *Aeromonas* wound infections include sepsis, necrotizing fasciitis, amputation, or death.

Aeromonas spp. grow well on most routine media that support the growth of Gram-negative bacteria including BAP, chocolate agar, and MAC. On MAC plates, the colonies will be gray, non-lactose fermenters. For wound, tissue, or blood specimens, these plates will generally show equivalent growth by 24 h. The colonies of most *Aeromonas* will be gray and large on BAP, MAC, and chocolate agar, with wide zone beta-hemolysis seen on BAP (Fig. 1.34a). For stool specimens, *Aeromonas* are generally identified from a BAP or MAC plate by performing a "sweep" oxidase test of confluent growth in the first/second zone of inoculation; *Aeromonas* are oxidase positive. A selective agar for *Aeromonas* and *Yersinia enterocolitica* (Cefsulodin-Irgasan-Novobiocin agar) can also be used for stool specimens. *Aeromonas* will appear pink with a white halo on this selective-differential agar, which is indistinguishable from *Yersinia*, thus requiring additional identification techniques. This agar is not typically used for other specimen types that may contain *Aeromonas*. Isolates of *Aeromonas* can be difficult to identify to the species level using biochemical reactions and automated biochemical-based identification systems. The most common gene sequencing used routinely in clinical microbiology laboratories is 16S rRNA gene sequencing (i.e., bacterial broad range sequencing), specifically the first 500 base pairs of the 5′ region of the gene. This target region can only identify *Aeromonas* to the genus level confidently. MALDI-TOF mass spectrometry is capable of identifying *Aeromonas* to the species level; however, this depends on the representative spectra in the database. Users should verify the representation of the genus in their database and ensure the targets are validated before reporting *Aeromonas* beyond the species level.

Aeromonas have generally predictable antimicrobial susceptibility profiles; however, some emerging resistance mechanisms have been described. *Aeromonas* spp. are generally resistant to ampicillin and penicillin/amoxicillin. The cephalosporins also show variable activity for the following drugs: cephalothin, cefazolin, cefoxitin (except *A. veronii biovar veronii* [100% susceptible]), cefuroxime, ceftriaxone, and cefotaxime. The following drugs typically show activity against *Aeromonas*; ciprofloxacin, gentamicin, amikacin, tobramycin, imipenem, and trimethoprim-sulfamethoxazole. The case above showed two different species with mostly identical susceptibility profiles though some variation was seen for multiple classes of drugs. Aeromonas have been described that contain cephalosporinases, TEM-24 plasmid-borne extended spectrum beta-lactamases, and NDM-1 carbapenemase.

Summary Points

- *Aeromonas* spp. are found naturally in aquatic ecosystems and are opportunistic pathogens to humans.
- Wound infections can be especially dangerous as *Aeromonas* is capable of causing necrotizing fasciitis, which can lead to severe tissue destruction, amputation, or death.

- Culture is the primary method of detection for *Aeromonas* in clinical specimens. *Aeromonas* grows robustly in aerobic and anaerobic conditions on all standard media capable of supporting Gram-negative organisms.
- Identification to the species level is difficult and may only be achievable using MALDI-TOF or broad range sequencing.

Selected References

Bandino JP, Hang A, Norton SA. The infectious and noninfectious dermatological consequences of flooding: a field manual for the responding provider. Am J Clin Dermatol. 2015;16(5):399–424.

Fernández-Bravo A, Figueras MJ. An update on the genus *Aeromonas*: taxonomy, pathogenicity, and infection. Micrororganisms. 2020;8:129.

Janda JM, Abbott SL. The genus *Aeromonas*: taxonomy, pathogenicity, and infection. Clin Micro Rev. 2010;23(1):35–73.

Marden JN, McClure EA, Beka L, Graf J. Host matters: medicinal leech digestive-tract symbionts and their pathogenic potential. Front Microbiol. 2016;13(7):1569.

Case 8

Marc Roger Couturier

A 63-year-old female with a past medical history of IgG kappa multiple myeloma presented to the emergency department of the local cancer hospital with nausea, vomiting, and diarrhea of several day duration. She sought medical care when the diarrhea became bloody. The patient had recently undergone high-dose chemotherapy and autologous peripheral blood stem cell transplant and was taking levofloxacin and fluconazole as prophylactic therapy. In the 6 days prior to arriving at the emergency department, she had profound neutropenic fever which peaked at 102 °F before admission, which was initially attributed to administration of a colony-stimulating factor drug. The patient initially received intravenous rehydration, infusion of platelets, and intravenous electrolyte replacement. After initial consultation in the emergency department, the patient was referred to the bone marrow transplant department for admission, blood cultures, infusion of broad-spectrum antibiotics (meropenem, vancomycin, and acyclovir), and further workup as deemed appropriate by the BMT team. The patient became afebrile in less than 24 h after administration of antibiotics.

The care team initially expected a bacterial source of infection that could include *Enterococcus* spp., *Staphylococcus aureus,* multiple possible members of the *Enterobacteriaceae,* or nosocomial pathogens such as *Pseudomonas*, *Acinetobacter*, and *Stenotrophomonas*. Fungemia with *Candida* spp. was less of a concern given the fluconazole prophylaxis though *Pichia kudriavzevii* (formerly *Candida krusei)* and

C. glabrata complex remained a possibility given the intrinsic and variable resistance (respectively). Twenty-four hours after admission and collection of blood cultures, the first aerobic bottle flagged positive for Gram-positive cocci in chains. The isolate was identified by phenotypic identification as a viridans group *Streptococcus* and considered a skin contaminant. The patient remained on meropenem and vancomycin at this time. Ninety-six hours after collection, the anaerobic bottle flagged positive for growth. Gram-variable rods were initially seen on Gram stain (Fig. 1.35) and upon re-stain, the stain was described as Gram-negative "snakey" rods (Fig. 1.36). The subculture of the anaerobic blood bottle failed to grow on solid media after 4 days of incubation on MacConkey agar, sheep blood agar (both incubated aerobically), and brucella agar (incubated anaerobically). In the time between the initial blood bottle flagging positive for growth and this time, the patient had significantly improved and had remained afebrile for 36 h. Eventually, an aliquot of the blood culture media was extracted for direct sequencing of the 5′ 500 base pairs of the 16S rRNA gene. *Leptotrichia hongkongensis* was identified with 100% sequence identity. Due to the lack of growth on solid media, antimicrobial susceptibility testing could not be performed. By the time the sequencing results were available to the clinical team, the patient was discharged on metronidazole for 10-day duration.

Discussion

The genus *Leptotrichia* is comprised of fastidious anaerobic, pencil-shaped/"snakey," Gram-negative rods (Figs. 1.35 and 1.36). These organisms are notoriously

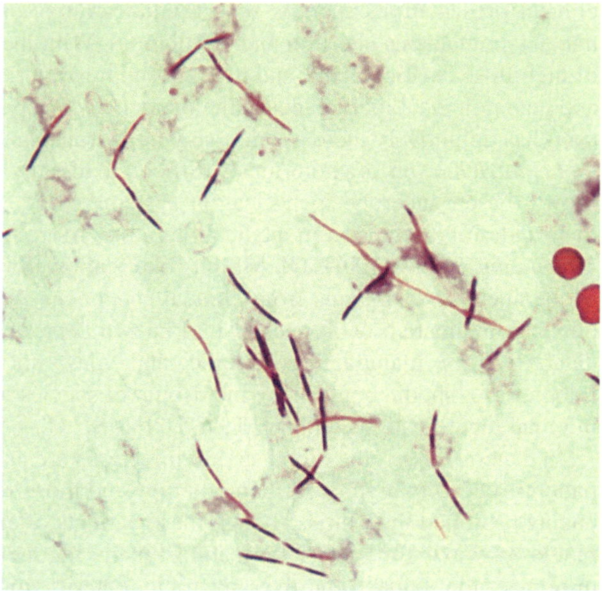

Fig. 1.35 Gram stain from positive blood bottle showing "snakey" or pencil-shaped Gram-variable rods (e.g., blunt on one pole and pointed at the opposite pole)

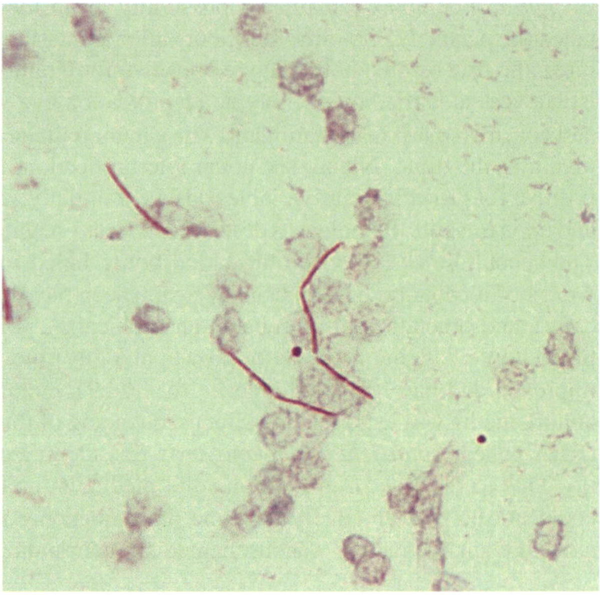

Fig. 1.36 Gram stain from second preparation showing Gram-negative rods with "snakey" or pencil-shaped morphology

challenging for microbiology laboratories to initially identify as they are known to retain crystal violet on initial Gram stain from primary sources, leading to initial misidentification as a non-spore forming Gram-positive rod. *Leptotrichia* represent normal resident microbiota of the human oral cavity, intestinal tract, and female genital tracts. Until the last decade, very few invasive infections with *Leptotrichia* had been reported in medical literature. The majority of cases were related to severe clinical presentations such as sepsis, endocarditis, and complications with pregnancy (spontaneous abortion and amnionitis). With the recent increase in adoption of high-dose chemotherapy and immunosuppressive regimens for both solid tumor and hematological malignancies, the incidence of opportunistic *Leptotrichia* infections has definitively increased in many large transplant and cancer centers.

Leptotrichia spp. are notoriously difficult to identify with classic biochemicals or growth-based anaerobe identification systems, which may have played a part in their underrepresentation in medical literature. More recent adoption of 16S rRNA sequencing and MALDI-TOF MS has likely allowed for more facile identification of members of this genus from clinical specimens. While DNA sequences in the public domain are broadly inclusive of known *Leptotrichia* spp., MALDI-TOF MS databases are manufacturer-specific and may not be as broadly inclusive. Laboratories should confirm the inclusivity of species within the genus before making final identifications with MALDI-TOF MS.

In general, anaerobic blood stream infections are an established complication to patients receiving high-dose chemotherapy and those who are subsequently experiencing profound mucositis. Though these patients often receive empiric antimicrobial prophylaxis for bacterial infections, many regimens are specifically chosen to minimize the impact on the anaerobic enteric microbiota to avoid inducing

Clostridioides difficile-associated disease. Levofloxacin, cefepime, and ceftazidime are examples of such drugs that have minimal anaerobic activity, and thus preserve the anaerobic enteric microbiota, but may allow for subsequent complications from opportunistic anaerobic infections. Though these broad families of anaerobic organisms may escape the intended spectrum of antimicrobial coverage, they are broadly susceptible to conventionally used empiric therapy for neutropenic fever. Limited antibiotic susceptibility data exist for *Leptotrichia;* however, the genus as a whole is susceptible to the majority of classes of antimicrobials. One notable exception is the species *L. trevisanii* which can be aerotolerant and thus display elevated MICs to metronidazole. Isolates may be tested for beta-lactamase activity as part of the routine antimicrobial susceptibility testing procedure; however, production of beta-lactamases is not reported. No acquired resistance determinants have been described in this genus to date. Most cases of *Leptotrichia* bacteremia/sepsis are quickly sterilized by empiric treatment for sepsis which includes anaerobic coverage.

The majority of clinical isolates of *Leptotrichia* are recovered from normally sterile sources such as blood, but they have also been recovered from normal physiological sites of colonization such as the oral cavity and genital tract. In these latter cases, the infection is typically an opportunistic infection due to a perturbation of the host's natural barriers (e.g., periodontal disease, dental surgery, vaginal wounds, premature rupture of membranes). In the clinical laboratory, *Leptotrichia* grows slowly on solid media and may show the most robust growth on brucella agar or chocolate agar. Growth on Columbia blood agar is generally supported but may result in slower growth and smaller colonies after 48 h of subculture from a positive blood bottle. Growth is typically seen in the anaerobic blood culture bottle; however subculture to solid media using an aerotolerance check plate can yield growth for *L. trevisanii*. Though the initial description of the species *L. hongkongensis* was recovered in solid culture, many reported cases with this species failed to grow on solid media after initial growth in blood culture bottles and required direct-from-specimen sequencing to yield an identification. In the case of blood cultures from patients with mucositis or other breakdowns of the mucosa, the cultures may often be polymicrobial. The most common organisms recovered with *Leptotrichia* are viridans group *Streptococcus*, *Enterococcus,* and occasionally yeast.

Summary Points

- *Leptotrichia* represents a normal component of the enteric microbiota that can occasionally cause opportunistic infections in patients with underlying medical conditions such as organ and bone marrow transplantation, pregnancy, and severe immunosuppression.
- Patients receiving high-dose chemotherapy with increased risk of mucositis are at increased risk of *Leptotrichia* bacteremia.
- *Leptotrichia* spp. are difficult to identify with routine biochemical reactions and require advanced molecular identification methods such as 16S rRNA sequencing or MALDI-TOF MS.

- The antimicrobial susceptibility profile of *Leptotrichia* is generally predictable with only some aerotolerant isolates displaying elevated MICs to metronidazole. Acquired resistance determinants is not currently a concern.

Selected References

Cho EH, Park KS, Yang M, Song DJ, Huh HJ, Ki CS, Lee NY. Laboratory identification of *Leptotrichia* species isolated from bacteremia patients at a single institution. Ann Lab Med. 2017;37(3):272–6.

Couturier MR, Slechta ES, Goulston C, Fisher MA, Hanson KE. *Leptotrichia* bacteremia in patients receiving high-dose chemotherapy. J Clin Microbiol. 2012;50(4):1228–32.

Eribe ER, Olsen I. *Leptotrichia* species in human infections. Anaerobe. 2008;14(3):131–7.

Schrimsher JM, McGuirk JP, Hinthorn DR. *Leptotrichia trevisanii* sepsis after bone marrow transplantation. Emerg Infect Dis. 2013;19(10):1690–1.

Woo PC, Wong SS, Teng JL, Leung KW, Ngan AH, Zhao DQ, Tse H, Lau SK, Yuen KY. *Leptotrichia hongkongensis* sp. nov., a novel *Leptotrichia* species with the oral cavity as its natural reservoir. J Zhejiang Univ Sci B. 2010;11(6):391–401.

Case 9

Megan H. Amerson-Brown and James J. Dunn

A previously healthy 7-week-old female, who was delivered vaginally at 36 weeks gestation, presented to the emergency department with a fever of 102 °F, decreased oral intake, irritability, sleepier than usual, decreased urine volume, one episode of emesis, and mild congestion. She had normal vitals, no abnormal movements, rash, diarrhea, or cough. Upon examination she had a positive Kernig sign and Brudzinski sign, indicating inflammation of the meninges, and slow capillary refill, taking 2–3 s, signifying hypoperfusion of the skin, increased systemic vascular resistance, and hypovolemia. A cerebral ultrasound revealed increased echogenicity lining the ventricles and the surface of the brain, suggestive of diffuse meningitis and ventriculitis. The differential diagnosis included bacteremia, UTI/urosepsis, common causes of late onset neonatal bacterial meningitis such as *Escherichia coli, Streptococcus agalactiae,* and *Listeria monocytogenes*; and possibly viral meningitis that might be caused by HSV or CMV. Pneumonia was considered less likely since there was a lack of respiratory symptoms. Initially, the patient was placed on empiric antibiotic therapy with IV ceftazidime and vancomycin and transferred to

Table 1.5 CSF laboratory test results

Laboratory test		Patient value	Reference range
CSF total WBC		5052 cells/μL	0–5 cells/μL
CSF WBC (%)	Neutrophils	73%	No reference range
	Lymphocytes	22%	No reference range
	Monocytes	5%	No reference range
CSF RBC		65 cells/μL	0–3 cells/μL
CSF glucose		<20 mg/dL	50–70% of serum glucose
CSF protein		565 mg/dL	15–45 mg/dL

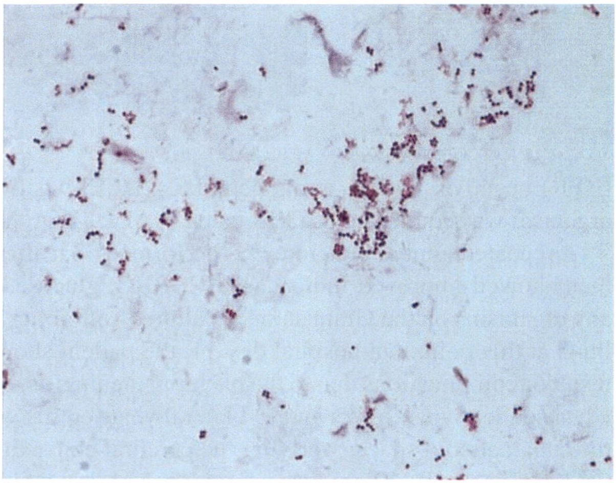

Fig. 1.37 Gram stain of a blood culture bottle growing *S. agalactiae* at 100× magnification

the PICU for further evaluation. Cerebral spinal fluid (CSF) collection by lumbar puncture was performed, and blood cultures were drawn.

The Gram stain from the first CSF revealed moderate Gram-positive cocci in pairs and chains, many WBCs and few RBCs. The CSF cell count, total protein and glucose were abnormal and consistent with an infectious etiology (Table 1.5). The CSF appeared light yellow with xanthochromia, the presence of bilirubin potentially indicating an acute subarachnoid hemorrhage or traumatic tap, and was slightly cloudy upon observation. On hospital day 2, the blood cultures flagged positive for Gram-positive cocci in pairs and chains (Fig. 1.37). The organism growing in the blood culture was identified as *S. agalactiae* (Group B Streptococcus or GBS) by a multiplex, molecular, microarray test system that can identify Gram-positive organisms commonly associated with bloodstream infections. GBS grew from both the CSF and blood culture on blood (BAP) and chocolate agar (CHOC). On BAP, the organism appeared small grayish white colonies approximately 1 mm in size with a narrow zone of beta-hemolysis (Fig. 1.38). On CHOC, the organism grew as small gray colonies. The organism had the following results when bench top biochemical testing was performed: catalase negative, pyrrolidonyl aminopeptidase

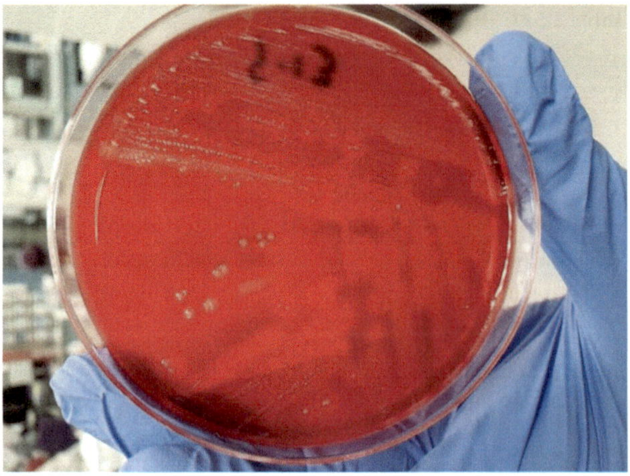

Fig. 1.38 *S. agalactiae* growing on blood agar as small gray colonies with a narrow zone of hemolysis

(PYR) negative, leucine aminopeptidase (LAP) positive. The identification of the organism was confirmed as GBS using MALDI-TOF MS.

A repeated lumbar puncture was performed 48 h after admission. The CSF specimen showed improved indices (WBC, RBC, glucose, protein) and did not reveal any organisms on the Gram stain or culture. Antibiotics were changed to IV penicillin G at this point. On hospital day 11, the patient showed signs of a seizure as the result of complications due to the bacterial meningitis. At this time, the MRI showed advanced leptomeningitis and left lateral ventriculitis with purulent material within the subarachnoid space overlying the cerebral convexities on the right greater than the left (Fig. 1.39). There was no evidence of drainable empyema. The patient was once again switched to ceftazidime and vancomycin but all cultures were negative. On hospital day 13, the patient was switched back to IV penicillin G that the patient received for an additional 29 days resulting in 42 days of total treatment. Prior to discharge the patient was given a hearing test in which no abnormal results were found. To date the patient has not presented with any additional sequelae.

Discussion

GBS is the leading cause of bacterial sepsis and meningitis in infants. GBS can be normal urogenital, gastrointestinal/rectal, or oropharyngeal flora in men and women of reproductive age. The organism can be transmitted from the mother to the fetus either in utero or during the birthing process. GBS neonatal infections can be divided into three categories: early onset that occurs during the first week of life, late onset that occurs after 1 week and before 3 months of life, and late-late onset that occurs after 3 months of life. Neonates with GBS bacteremia and/or meningitis are prone to adverse outcomes including mortality rates as high as 50% and

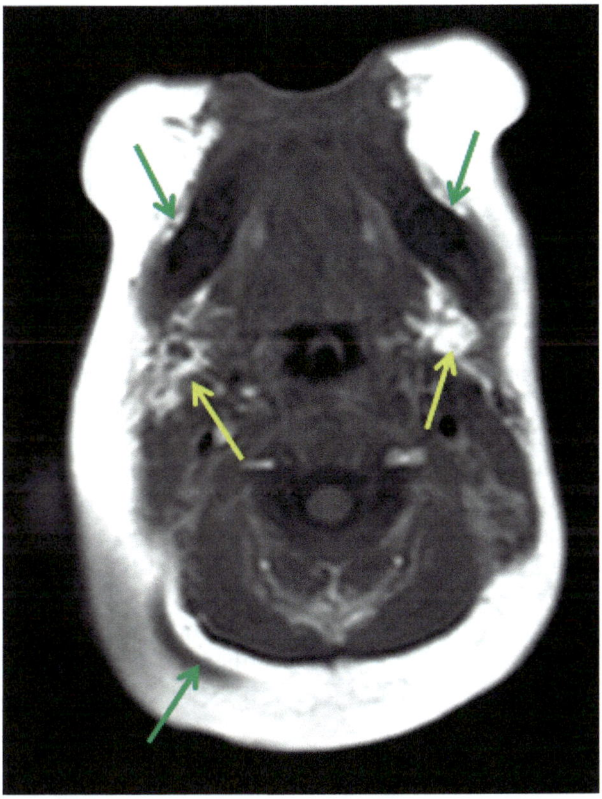

Fig. 1.39 MRI showing intracranial volume loss, purulent material within the subarachnoid spaces (green arrows) and enhancement of the leptomeninges (yellow arrows)

neurological sequelae. Despite efforts in prevention, resulting in a significant decrease of infections, GBS sepsis and meningitis is still a concern.

Approximately 25% of women of reproductive age are colonized with GBS. The US Centers for Disease Control and Prevention (CDC) and the American Society for Microbiology (ASM) have both issued guidance around screening and prenatal testing for GBS. Per current recommendations, prenatal screening for GBS colonization is performed between 36 and 38 weeks of pregnancy by collection of a single vaginal-rectal swab for laboratory testing. The predictive value of GBS prenatal testing decreases significantly when collected greater than 5 weeks from delivery. Bacterial culture methods using selective media are preferred over molecular testing for prenatal GBS screening. If molecular testing is used, the specimen should be inoculated into selective enrichment broth that allows for the growth of GBS but inhibits growth of other vaginal microbiota. The molecular test should be performed from the enrichment broth 18–24 h after inoculation. Rapid molecular testing for GBS is available for women in active labor with no previous screening results; however, these tests without broth enrichment have a high false-negative rate.

Efforts to reduce early onset GBS can be prevented by administering intravenous antibiotic prophylaxis prior to delivery in colonized women who had a positive GBS screening test. Late-late onset disease is likely caused by GBS transmission post-delivery either from the mother or healthcare setting to the neonate. There is currently no effective approach for the prevention of late-onset of GBS infection in neonates.

GBS infections in adults are typically observed in elderly or immunocompromised patients with underlying conditions of alcoholism, diabetes mellitus, or cancer. Recent hospitalization can also be considered a risk factor. Approximately one in four cases in nonpregnant adults are nosocomially acquired. Clinical presentation in adults most commonly consists of bacteremia without a primary focus, osteoarticular infections, focal infections, and skin and soft tissue infections.

Recovery of GBS in routine culture is good given an optimal sample and appropriate transport conditions. Colonies will grow on BAP, CHOC, Colistin nalidixic acid agar (CNA) and broth media including blood culture bottles. Colonies usually appear gray with a small zone of hemolysis on blood containing media. On rare occasions, non-hemolytic isolates of GBS have been isolated and are sometimes confused with *Enterococcus* species. Benchtop biochemical can be used to easily differentiate GBS from *Enterococcus*, *Staphylococcus*, and other *Streptococcus* species. GBS will be catalase negative, LAP positive, PYR negative, and hippurate hydrolysis positive. An agglutination test for Lancefield groups of beta hemolytic *Streptococcus* will agglutinate only with Group B reagents (Fig. 1.40). A CAMP test, when performed with GBS, will show enhanced hemolysis at the *Staphylococcus aureus* and GBS junction. Identification and reporting of GBS from all sources is

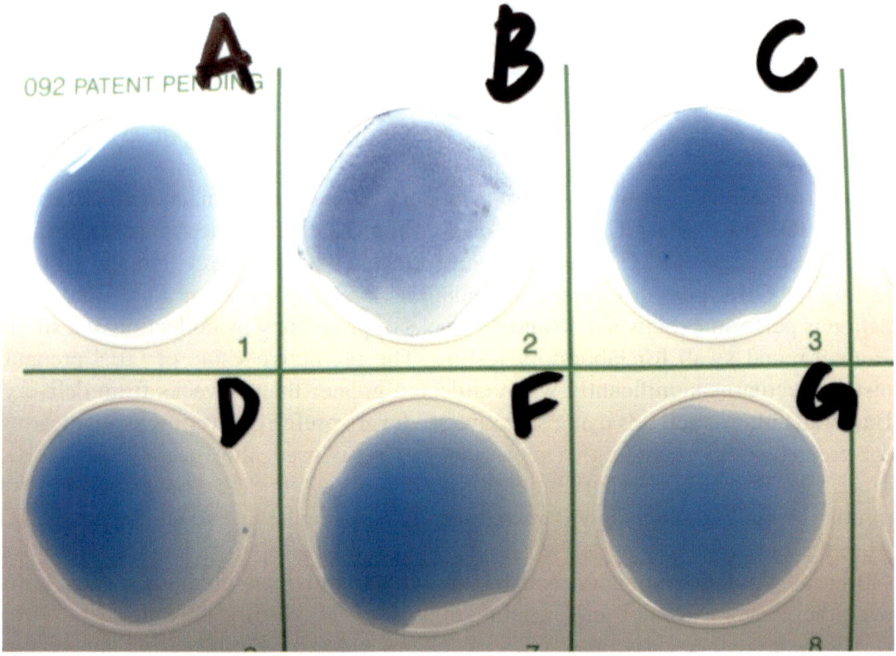

Fig. 1.40 Streptococcus latex agglutination test. Agglutination is based on surface antigens that correspond with the Lancefield group. The organism caused agglutination with the Group B reagents. All other reagents had no agglutination. This organism is Lancet group B Streptococcus

needed to identify extra-vaginal infections. Most molecular-based diagnostic testing panels for identification of organisms from blood cultures bottles include GBS along with other common streptococcal species, such as *S. pneumoniae*.

Routine antimicrobial susceptibility testing is not recommended for GBS isolates recovered from prenatal screening for GBS colonization. Penicillin and ampicillin are the drugs of choice for the treatment of GBS intrapartum antibiotic prophylaxis. Cefazolin can also be used for treatment in penicillin-allergic persons who are at low risk for penicillin anaphylaxis. In cases where the person has a severe penicillin allergy erythromycin and clindamycin should be tested and only clindamycin should be reported for isolates obtained from vaginal-rectal swabs in cases of antepartum screening. Inducible macrolide resistance should be screened for to accurately report clindamycin susceptibility. Isolates of GBS non-susceptible to penicillin and ampicillin are extremely rare. If susceptibility testing is performed, Clinical and Laboratory Standards Institute (CLSI) guidelines should be followed in order to provide a susceptibility interpretation based on the determined minimum inhibitory concentration value or a zone of inhibition when using disk diffusion.

Summary Points

- GBS can colonize the urogenital tract and can be detected in approximately 25% of women of reproductive age. GBS can cause severe invasive disease in neonates during the first year of life and in the elderly or immunocompromised persons.
- CDC recommends screening for GBS in all pregnant women at 36–38 weeks of gestation and administration of IV antibiotics during labor if screening is positive.
- GBS is easily isolated by routine culture methods.
- Antimicrobial susceptibility testing for GBS isolates obtained from antenatal screening for colonization is not routinely recommended due to rates of penicillin resistance being low, however if susceptibility testing is performed due to a severe penicillin allergy, inducible clindamycin resistance must be determined.

Selected References

Filkins L, Hauser JR, Robinson-Dunn B, Tibbetts R, Boyanton BL, Revell P. American Society for Microbiology provides 2020 guidelines for detection and identification of group B *Streptococcus*. J Clin Microbiol. 2020;59(1):e01230-20.

Francois Watkins LK, McGee L, Schrag SJ, Beall B, Jain JH, Pondo T, Farley MM, Harrison LH, Zansky SM, Baumbach J, Lynfield R, Snippes Vagnone P, Miller LA, Schaffner W, Thomas AR, Watt JP, Petit S, Langley GE. Epidemiology of invasive group B streptococcal infections among nonpregnant adults in the United States, 2008–2016. JAMA Intern Med. 2019;179(4):479–88.

Heath PT, Jardine LA. Neonatal infections: group B streptococcus. BMJ Clin Evid. 2014;2014:0323.

Nanduri SA, Petit S, Smelser C, Apostol M, Alden NB, Harrison LH, Lynfield R, Vagnone PS, Burzlaff K, Spina NL, Dufort EM, Schaffner W, Thomas AR, Farley MM, Jain JH, Pondo T, McGee L, Beall BW, Schrag SJ. Epidemiology of invasive early-onset and late-onset group B streptococcal disease in the United States, 2006 to 2015: multistate laboratory and population-based surveillance. JAMA Pediatr. 2019;173(3):224–33.

Case 10

James J. Dunn

A 7-month-old male was brought to the emergency room because he had not been bearing weight on his right leg for about a week. He also recently had some low-grade fevers, congestion, and a mild cough about 2 weeks prior to presentation. The child had been born at term with no complications. Upon examination, the child had full range of motion of the right lower leg but would not bear weight on it when held upright. There appeared to be some mild swelling over the ankle area and possibly some tenderness to pressure.

Initial laboratory testing revealed an elevated white blood cell (WBC) count of 22,200 cells/μL (reference range, 5.98–13.51 × 10^3 cells/μL) with 43% neutrophils (reference range, 16.9–74%), 45% lymphocytes (reference range, 26–79.9%), and 11% monocytes (reference range, 3.8–13.4%). The erythrocyte sedimentation rate was elevated at 31 mm/h (reference range, <1–20 mm/h) and the C-reactive protein level was also high at 6.5 mg/dL (reference range, <1.0 mg/dL), indicating a systemic inflammatory process. Blood cultures were obtained and the patient was admitted and started on broad-spectrum antibiotic therapy. An MRI of the right leg showed bifocal tibial osteomyelitis with a distal tibial abscess crossing the growth plate and an ankle joint effusion (Fig. 1.41).

The child was evaluated by the orthopedic surgery service, and it was determined that surgical incision and drainage of the right distal tibia and ankle be performed. In the operating room, the surgeon noted purulent material which extended down the physis of the tibia into the epiphysis and ankle joint. The intraosseous and ankle joint fluid was collected and sent to the laboratory for bacterial, AFB, and fungal culture. In addition to inoculation onto agar plates, a portion of the joint fluid was inoculated into an aerobic blood culture bottle and incubated on the continuously monitored blood culture instrument. The ankle joint fluid was also subjected to real-time PCR testing for *Kingella kingae*.

Results of microbiologic testing showed no organisms present in the Gram stain of the ankle joint fluid and few WBC. No organisms were detected in blood cultures obtained prior to admission and cultures of the joint fluid for bacteria (both agar and bottle), AFB, and fungi showed no growth. Only the real-time PCR test for *K. kingae* was positive (Fig. 1.42).

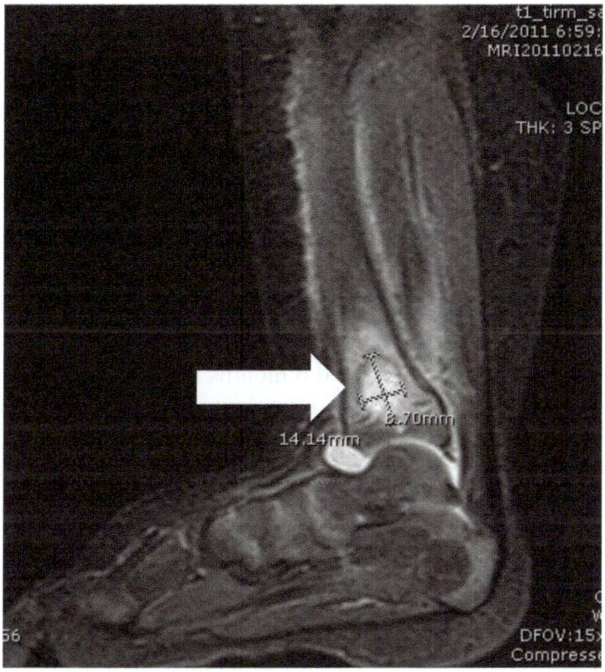

Fig. 1.41 MRI imaging of the right lower leg of a 7-month-old male unable to bear weight on the limb. The image shows bifocal tibial osteomyelitis with a distal tibial abscess crossing the growth plate and an ankle joint effusion (arrow)

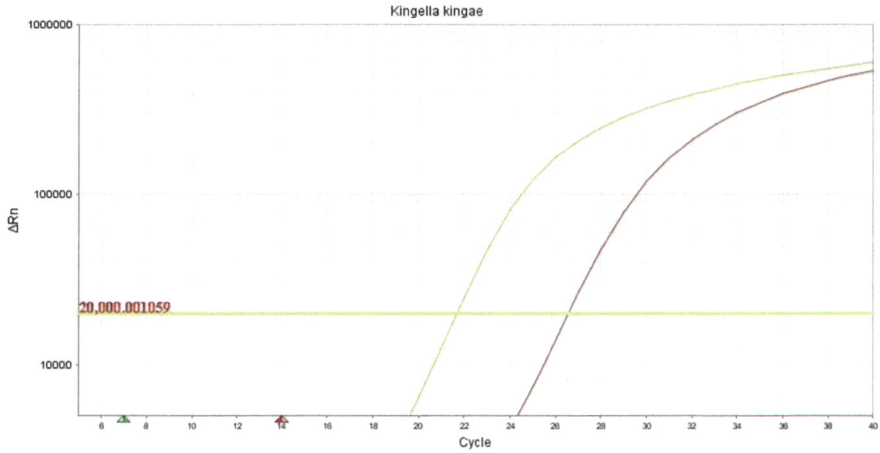

Fig. 1.42 Real-time PCR amplification curve for detection of *Kingella kingae* in joint fluid. The horizontal green bar represents the threshold above which amplification indicates a positive result. The yellow curve represents the real-time amplification and detection of *K. kingae* from the patient's synovial fluid. The red line is amplification of the positive control in a separate reaction

Discussion

Clinical disease syndromes caused by *K. kingae* primarily include bacteremia, bone and joint (osteoarticular) infections, and endocarditis. The vast majority of osteoarticular infections occur in children less than 4 years of age whereas endocarditis is more commonly seen in adult patients. Most cases of septic arthritis are monoarticular and most involve the knee, hip, or ankle. Oftentimes, the joint is aspirated and purulent fluid is obtained which contains predominantly neutrophils. Osteomyelitis due to *K. kingae* infection usually involves the long bones such as the femur and tibia. Other skeletal infections caused by *K. kingae* include spondylodiskitis, dactylitis, and tenosynovitis.

In young children aged 6 months to 4 years, *K. kingae* may intermittently colonize the upper respiratory tract; up to 9–12% in children 12–24 months of age. It is postulated that in some instances viral upper respiratory tract illness may precede invasive infection with this organism. In these cases, the organism colonizes the posterior pharynx and virus-induced damage to the respiratory mucosa facilitates *K. kingae* translocation across the pharyngeal epithelium to enter the bloodstream. Once in the bloodstream, the organism may disseminate and seed joints, bone or heart tissue.

K. kingae is Gram-negative facultative anaerobic bacteria that is arranged in pairs or short chains of bacilli on Gram stain (Fig. 1.43). It grows optimally at 37 °C in a 5–10% CO_2-enhanced atmosphere on blood and chocolate agar. No growth is observed on MacConkey agar. On blood agar colonies exhibit a narrow zone of β-hemolysis (Fig. 1.43). The organism is notoriously fastidious and difficult to recover using solid media culture-based methods. Recovery can be improved by inoculating specimens into an aerobic blood culture bottle with incubation on an automated detection instrument similar to the process used for blood culture. If growth of the organism occurs, it can be readily identified using MALDI-TOF mass spectrometry or commercially available biochemical panels.

The use of nucleic acid amplification testing (e.g., PCR) has dramatically increased the number of identified cases of *K. kingae* infections and has shown that it can be detected in a large number of culture-negative joint fluid and bone samples from patients with suspected osteoarticular infections. Some institutions using PCR techniques now report that *K. kingae* is responsible for the majority of cases of septic arthritis in children 6–36 months of age; even more than *S. aureus* or *S. pyogenes*. In this case, there was no growth of *K. kingae* on solid media when directly plated with the ankle joint fluid or in the blood culture bottle inoculated with the same material. Infection with *K. kingae* was identified only by the use of PCR on the joint fluid, thus directing the healthcare provider to administer the appropriate antibiotic treatment for this organism in a timely manner.

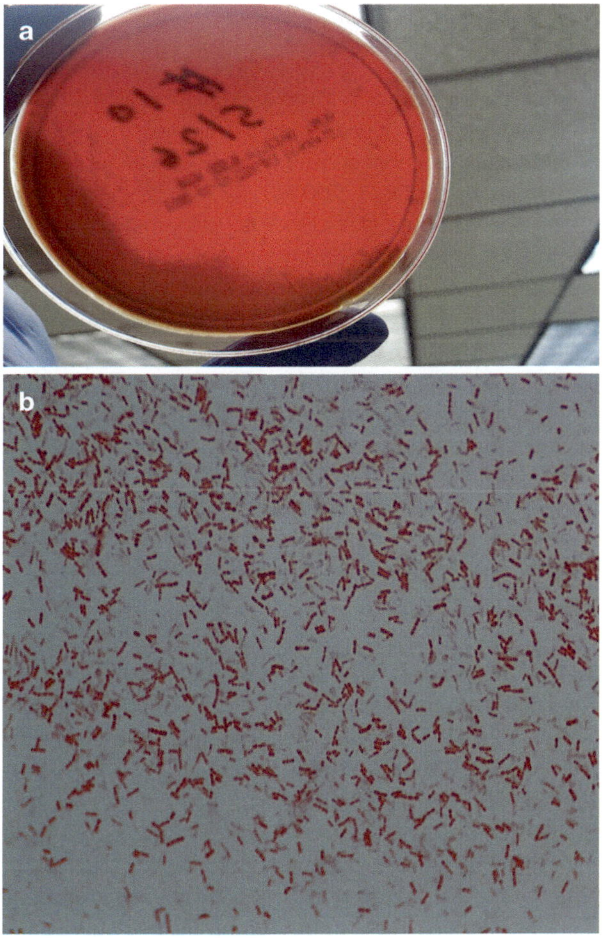

Fig. 1.43 Blood agar plate showing growth of *Kingella kingae* with a small zone of β-hemolysis around the colonies (**a**). Gram stain of *Kingella kingae* from colonies isolated on blood agar (**b**)

Summary Points

- *K. kingae* is a Gram-negative bacilli that is an important pathogen responsible for a large number of bone and joint infections in children less than 4 years of age.
- Although growth of *K. kingae* is possible on routine bacteriologic media, its fastidious nature often results in negative cultures of bone and joint fluid specimens.
- New and sensitive molecular diagnostic techniques such as real-time PCR for *K. kingae* are imperative to identify the causative agent of bone and joint infections in children.
- Rapid and accurate diagnosis of *K. kingae* osteoarticular infection allows for prompt administration of appropriate antibiotics; treatment distinctly different than ones used for other pathogens known to cause bone and joint infections.

Selected References

Chometon S, Benito Y, Chaker M, Boisset S, Ploton C, Berard J, Vandenesch F, Freydiere AM. Specific real-time polymerase chain reaction places *Kingella kingae* as the most common cause of osteoarticular infections in young children. Pediatr Infect Dis J. 2007;26:377–81.

Wong M, Williams N, Cooper C. Systematic review of *Kingella kingae* musculoskeletal infection in children: epidemiology, impact and management strategies. Pediatr Health Med Ther. 2020;11:73–84.

Yagupsky P. *Kingella kingae*: carriage, transmission, and disease. Clin Microbiol Rev. 2015;28:54–79.

Yagupsky P. Diagnosing *Kingella kingae* infections in infants and young children. Expert Rev Anti Infect Ther. 2017;15:925–34.

Case 11

Marc Roger Couturier

A 49-year-old male with a past medical history of type 2 diabetes, hypersensitivity lung disease, and obstructive sleep apnea presented to the emergency department in Salt Lake City, UT with a chief complaint of shortness of breath and a nonproductive cough of 6-week duration. He had previously been in his usual state of health until he developed the aforementioned progressive shortness of breath and nonproductive cough which initially were accompanied by pleuritic chest pain and poor appetite. Early in his clinical course, he experienced pharyngitis and nasal congestion, but these symptoms were brief in duration and resolved spontaneously. Throughout the course of his symptoms he denied any fevers, chills, nausea, vomiting, diarrhea, or any sick contacts. He was tested for SARS-CoV-2 on numerous occasions during his acute illness but was repeatedly negative. He was fully vaccinated for influenza and SARS-CoV-2. During this time, he had been seen at a local emergency department in Jackson, Wyoming on two different occasions without a definitive diagnosis. At his most recent hospital encounter, he was prescribed prednisone and azithromycin for "walking pneumonia" though no chest X-ray was performed. He endorsed moderate improvement of his symptoms while on prednisone; however, the symptoms worsened once he completed the taper. In the week preceding his hospital visit in Utah, he reported intermittent dizziness but denied any falls or difficulty with balance. The patient also denied any numbness or tingling in his extremities during this period. A complete metabolic panel was performed which was only significant for elevated bilirubin and glucose (Table 1.6). His complete blood count was remarkable for thrombocytopenia and elevated neutrophils (Table 1.7). A chest X-ray was unremarkable and a chest CT was noted for mild thickening in both lower lobes and no signs of pulmonary emboli or infiltrates. Small thin-walled pulmonary cysts were documented, but they were seen on a

Table 1.6 Complete metabolic panel results at admission

Analyte	Value	Reference range
Sodium	**128 ↓**	136–144 mmol/L
Potassium	4.3	3.3–5.0 mmol/L
Chloride	**97 ↓**	102–110 mmol/L
Carbon dioxide	22	20–26 mmol/L
BUN	19	8–24 mg/dL
Creatinine	0.8	0.72–1.25 mg/dL
Glucose	**318 ↑**	64–128 mg/dL
Calcium	9.4	8.4–10.5 mg/dL
Protein, Total	7.9	6.5–8.4 g/dL
Albumin	3.6	3.5–5.0 g/dL
Bilirubin, Total	**2.3 ↑**	0.2–1.4 mg/dL
Alkaline phosphatase	65	38–126 U/L
AST	25	16–40 U/L
ALT	32	5–60 U/L
Anion gap	9	8–14 mmol/L

The bold values indicates the results that are resistant or abnormal

Table 1.7 Complete blood cell count at admission

Analyte	Value	Reference range
WBC	**12.51 ↑**	4.30–11.30 k/μL
RBC	5.42	4.70–6.14 M/μL
HGB	16.2	14.8–17.8 g/dL
Hct	46.8	44.2–53.0%
MCV	86.3	81.2–96.6 fL
MCH	29.9	25.8–33.1 pg
MCHC	34.6	31.9–35.2 g/dL
RDW	13.4	11.5–15.3%
Platelet	**92 ↓**	159–439 k/μL
Neutrophil %	70.1	39.4–72.5%
Lymphocyte %	20.1	17.6–49.6%
Monocyte %	7.4	4.1–12.4%
Eosinophil %	0.6	0.4–6.7%
Basophil %	0.4	0.3–1.4%
Neutrophil #	**8.76 ↑**	2.00–7.40 k/μL
Lymphocyte #	2.52	1.30–3.60 k/μL
Monocyte #	0.93	0.30–1.00 k/μL
Eosinophil #	0.07	0.00–0.50 k/μL
Basophil #	0.05	0.00–0.10 k/μL

The bold values indicates the results that are resistant or abnormal

previous scan several years earlier and deemed benign. Nonetheless, infectious diseases that could explain these cysts were considered (e.g., tuberculosis, pneumocystosis, coccidiomycosis).

The patient's social history was significant for cigarette and e-cigarette use for a cumulative time greater than 25 years. He currently works as a rancher in western

Wyoming and primarily raises cattle and sheep. He has multiple sheep dogs, including two wolf-husky hybrids that live outdoors but sleep with the patient while driving stock on rangelands. He endorsed eating meat from both mutton and older bulls on a regular basis, however stated that he stews the meat and cooks it well. The patient denied consuming raw milk or cheese. He spends a great deal of time outdoors and has experienced multiple fly, mosquito, and tick bites recently. The patient also has significant exposures to bovine and ovine products of conception, including within the month prior to his symptom onset. Consultation with infectious disease was sought and a comprehensive list of tests were ordered (Table 1.8). The only positive results were Q fever phase II IgG and IgM, and phase I IgM serology (Table 1.8).

The patient was diagnosed with acute Q fever and treated with a 14-day course of doxycycline. He was instructed to have follow-up serology drawn at 3, 6, and 12 months from completion of therapy. The patient's respiratory symptoms resolved rapidly with therapy, however the fatigue persisted, but was slowly improving. His thrombocytopenia also resolved with therapy.

Discussion

Coxiella burnetii is a small obligate intracellular Gram-negative bacterium that causes Q fever. The organism has two morphological forms, termed the large cell variant (LCV) and the small cell variant (SCV). The LCV is the metabolically active, replicative form that resides inside of a parasitophorous vacuole within the infected host cell. The SCV is akin to endospore stages of Gram-positive organisms, and are metabolically inactive, typically in response to adverse conditions and nutrient depletion. The SCV is responsible for transmission to new hosts and is found in the environment in dust, dirt, water sources, and within the vacuoles of free-living amoebae. Ticks within the Ixodidae and Argasidae families have been known to harbor *C. burnetii;* however, they are not definitively known to be competent vectors for infection. Many vertebrates can be infected (symptomatically or not) by *C. burnetii* including many ungulates (including bovine, ovine, porcine, caprine, and cervine species), felines, canines, lagomorphs, and avians. Most of these animals do not show any clinical signs and symptoms of infection. Exposure to products of conception from the aforementioned birthing animals is the most common route of infection to humans. The ubiquitous distribution of *C. burnetii*, worldwide practice of animal husbandry, combined with the extremely low infectious dose (~1–10 organisms) is the reason that infections with this organism are more frequent than many other highly pathogenic select agents.

C. burnetii temporally exists in two forms of lipopolysaccharide (LPS) expression: phase I (virulent) and phase II (avirulent). The phase II form induces apoptosis

Table 1.8 Infectious disease testing results collected at admission

Assay	Source	Analyte/Target	Result
Respiratory pathogen panel	Nasopharyngeal swab	Influenza A & B, RSV, SARS-CoV2, Parainfluenza 1–4, human metapneumovirus, rhinovirus, adenovirus, *Chlamydia pneumoniae, Mycoplasma pneumoniae, Bordetella pertussis*	Not detected
Pneumocystis direct fluorescent antibody stain	Sputum	*Pneumocystis jirovecii*	Negative
Legionella antigen	Urine	*Legionella pneumophila* serogroup 1	Negative
Streptococcus pneumoniae antigen	Urine	*Streptococcus pneumoniae*	Negative
Blood culture	Blood	N/A	Negative
Respiratory bacterial culture	Sputum	N/A	1+ Normal upper respiratory flora
Interferon gamma release assay	Whole blood	*Mycobacterium tuberculosis*	Negative
Tuberculosis PCR	Sputum	*Mycobacterium tuberculosis*, rifampin resistance detection	Not detected
Acid-fast bacilli smear and culture (X3)	Sputum	N/A	Negative (×3)
Brucella agglutination antibodies	Serum	Total antibodies: IgG, IgM, IgA combined	Negative
Rickettsia antibodies	Serum	IgM	Negative
	Serum	IgG	Negative
Coccidioides antibody	Serum	IgG	Negative
	Serum	IgM	Negative
Echinococcus antibody	Serum	IgG	Negative
Q fever phase I antibody	Serum	IgM	**1:4096***
	Serum	IgG	Negative
Q fever phase II antibody	Serum	IgM	**1:1024***
	Serum	IgG	**1:32768***

of the infected host cells and results in the acute symptomatic phase of illness characterized by localized inflammation and resulting symptoms. Eventually, the patient recovers either from sterilizing or non-sterilizing immunity. In the case of non-sterilizing immunity, a dormant form of the bacteria (SCV in phase I described below) can reactivate later, causing chronic Q fever complications. This risk is significantly lower when antibiotic therapy complements the immune response. Acute

Q fever typically presents as a mild pneumonia with thrombocytopenia, but the overall symptoms are neither specific nor highly predictive. Many patients experience uncharacteristically severe headaches, which can be more predictive for Q fever when paired with the aforementioned symptoms. Many clinicians may not be familiar with this organism, and thus additional cases may go undiagnosed simply because the organism cannot be diagnosed by routine culture and requires indirect (serology) or culture-independent detection techniques (NAAT) that are rarely ordered. Acute Q fever respiratory disease is commonly mistaken for a "walking pneumonia" or viral respiratory infection, and therefore untreated (or treated inadequately with a macrolide).

The phase I form of LPS functions as an immunological evasion mechanism, evading the dendritic cell response. When phase I bacteria encounter the innate immune response, they functionally bypass the complement response and infect host cells with essentially no immune recognition. Virulence factors are expressed that allow disruption of phagosome maturation, and the organism can survive as LCVs within the vacuole. *C. burnetii* can delay or block apoptosis, allowing unregulated replication of the cell. Once the organism has replicated to a critical mass within the vacuole, the LCVs become SCVs and the organism goes dormant within the immortalized/infected cell leading to a chronic infection state. Chronic Q fever can present with endocarditis (native or non-native valve), pericarditis, myocarditis, splenic rupture, hepatitis with granulomas, cholecystitis, and rarely meningitis and meningoencephalitis.

The primary method for diagnosis for Q fever is IgG serology. Specific antibodies for phase I and phase II are detected through a tittered assay (e.g., indirect immunofluorescent antibody detection [IFA]) with excellent specificity and sensitivity. Detection of phase I and phase II antibodies allow the stage of infection to be predicted based on the relative increase and ratio of the titers. Phase II IgG titers showing a four-fold increase between acute and convalescent specimens (collected 3–6 weeks apart) is diagnostic for acute Q fever. Antibodies against both phases are often detected in many Q fever patients, sometimes due to previous infections/exposures or the current infection. If acute Q fever is detected and treated early, a patient may not produce phase I IgG antibodies. Chronic Q fever will always have phase II IgG detected since the infection always begins with immune recognition of phase II LPS before progressing to phase I. In chronic Q fever, the phase I titers are typically much higher than phase II. Though no single titer criteria has been established for chronic Q fever, a titer ≥1:1024 is generally necessary to establish a diagnosis of chronic Q fever. IgM antibodies do not have the same specificity as IgG and should not be interpreted by themselves to establish a diagnosis. In fact, like many other intracellular pathogens, IgM is often detected around the same time as IgG, limiting the diagnostic utility for this analyte. In general, if an IgM titer is detected in the absence of IgG, the result should be interpreted as presumptive positive, and an additional specimen should be collected in 10–14 days to confirm the corresponding IgG. High-titer phase II IgM can generate low-titer false-positive phase I IgM results. Importantly, false-positive IgG results have also been reported in patients with *Bartonella* infections.

NAAT testing can be useful for very early disease when testing blood specimens. The blood should be collected during the first 2 weeks of symptoms, when the bacteremia is at its highest concentration. NAAT is significantly more sensitive than serology in these first 2 weeks; however, beyond the third-week serology is more sensitive. The sensitivity of the NAAT can be increased if the buffy coat is tested since the infected cells are concentrated in this specimen. Serum is not recommended since there is no cellularity in the specimen, and cell-free nucleic acid is all that will be detected. For tissue specimens from chronic Q fever (e.g., excised—heart, liver, or splenic tissue), NAAT can be a valuable direct detection method. Paradoxically, patients with phase I IgG titers between 1:800 and 1:6400 are typically positive by NAAT for chronic Q fever, whereas lower or higher titers are typically NAAT-negative. NAAT assays targeting a multicopy gene are significantly more sensitive but may be less specific. Serology should be used to help support NAAT-positive results.

The treatment for acute Q fever typically consists of 14 days of doxycycline followed by serial testing of IgG titers. Phase II IgG titers should decline over a year and stay low or become undetectable. If a patient develops phase I titers at any point during the follow-up monitoring, treatment for chronic Q fever should be considered. For chronic Q fever, treatment involves a combination of doxycycline and hydroxychloroquine for a year or longer. Phase I and phase II titers may remain elevated for years after therapy and may never become undetectable. The titers should stay the same over time (± two-fold titer differences between specimens over time).

Summary Points

- *Coxiella burnetii* is an obligate intracellular Gram-negative bacterium that cannot be cultured using routine laboratory methods.
- *C. burnetii* causes Q fever in humans and can be acquired from the environment or through interactions with infected animals.
- Q fever can present as an acute pneumonia form or as a chronic infection, commonly as an infection of large organs such as the heart.
- Serology is the main detection method for Q fever, with phase I and phase II IgG antibodies being the most reliable analytes.
- Phase II IgG predominance and four-fold increase is considered diagnostic for acute Q fever. Phase I elevated titers with predominance over phase II is generally supportive of chronic Q fever.

Selected References

Centers for Disease Control and Prevention. Q fever. https://www.cdc.gov/qfever/index.html. Accessed 6 Mar 2022.

Kersh GJ and Bleeker-Rovers, CP. Chapter 68. *Coxiella*. In: Carroll KC, Pfaller MA, editors. Manual of clinical microbiology. 12th ed. Washington, DC: ASM Press; 2019. p. 1180–6.

Million M, Raoult D. Recent advances in the study of Q fever epidemiology, diagnosis and management. J Infect. 2015;71 Suppl 1:S2–9.

Sahu R, Rawool DB, Vinod VK, Malik SVS, Barbuddhe SB. Current approaches for the detection of Coxiella burnetii infection in humans and animals. J Microbiol Methods. 2020;179:106087.

Chapter 2
Virology Cases

James J. Dunn, Angela Ma, Cameron A. Brown, and Marc Roger Couturier

Overview of Diagnostic Virology Methods

Virology is one of the newer diagnostic microbiology subspecialties, as viruses and the means to cultivate them were discovered well after the characterization of bacteria and other microorganisms. During its relatively short history, clinical virology has seen many changes. Many new viruses and viral diseases have been identified. Much has been learned about the properties of viruses and the pathogenesis of viral infection. Diagnostic virology has evolved from an esoteric into an important routine laboratory service as effective antiviral agents have become available and testing has become faster and more accurate. To a large extent, this is the driving force

J. J. Dunn (✉)
Department of Pathology, Texas Children's Hospital, Houston, TX, USA

Department of Pathology & Immunology, Baylor College of Medicine, Houston, TX, USA
e-mail: jjdunn@texaschildrens.org

A. Ma
Public Health Ontario, Department of Laboratory Medicine and Pathobiology, University of Toronto, Toronto, ON, Canada
e-mail: Angela.Ma@oahpp.ca

C. A. Brown
Departments of Pathology & Immunology and Pediatrics, Baylor College of Medicine, Houston, TX, USA
e-mail: Cameron.Brown@bcm.edu

M. R. Couturier
Department of Pathology, University of Utah/ARUP Laboratories, Salt Lake City, UT, USA
e-mail: Marc.Couturier@aruplab.com

J. J. Dunn et al., *Challenging Cases in Diagnostic Clinical Microbiology*,
https://doi.org/10.1007/978-3-031-65092-5_2

behind the evolution of diagnostic virology. The clinical virology laboratory is now able to provide diagnoses that impact patient management and treatment decisions and public health interventions.

Viruses are infectious agents that are dependent on host energy sources for propagation. As such, viral replication must occur within a living cell. Viruses have a more simplistic structure than cellular organisms, consisting of a nucleic acid genome surrounded by a protein capsid, and in some cases an envelope consisting of a lipid bilayer with associated proteins. The complete unit is often referred to as a virion. Unlike bacteria, which reproduce by binary fission, the assembly of nucleic acid and protein components produces new virions. The replication strategy of viruses depends on the nature of their genome, which consists of either RNA or DNA. For basic viral replication, the virus must enter the cell, where the viral genome is replicated and directs the synthesis of macromolecules (mRNA and protein). New (progeny) virions are then formed inside the cell from the replicated genome and macromolecules. Finally, infectious progeny virions must then exit the cell and come into contact with a susceptible host or host cell to remain infectious. A wide variety of living creatures are susceptible to viral infections, including animals, plants, fungi, and bacteria, but this section will focus on methods to diagnose viral infections in humans.

Specimens

A common phrase used in the diagnostic microbiology laboratory "garbage in; garbage out" exemplifies in clear (if not overstated) fashion the importance of a quality specimen for accurate laboratory results. Optimal performance of diagnostic virology tests requires the sampling of the appropriate anatomic site, and when viral titers are highest (usually early in the disease). Since viruses are obligate intracellular parasites, specimens collected from certain sources (e.g., skin lesions, respiratory tract) generally must contain cells for optimal test sensitivity. Most viruses are extremely fragile outside of their normal host environment, and lose viability rapidly when stored desiccated or at room temperature for short periods of time, or even under optimal conditions beyond a couple days. Rigorous adherence to proper handling and transport of specimens is also necessary for optimal and accurate test performance.

To recognize the importance of timing of specimen collection, one must understand the basic pathogenesis of viral infections. Symptomatic infections generally consist of the following phases: incubation, acute (including prodrome during early acute), and convalescence (during which infection resolves, becomes latent, or active/subclinical). The duration of these phases varies greatly among diseases. During the incubation phase, infection of susceptible host cells begins and viral titers increase greatly. Clinical symptoms are absent or subclinical during this stage. Signs and symptoms develop during the acute phase, either gradually and mildly (prodrome), or abruptly. Generally, the level and duration of virus shedding depend on the virus, infected organ or organ system, and host factors such as age and immune status. Specimens for direct viral testing are generally best collected shortly after onset of symptoms (acute phase of illness), when virus titers are highest. For many viral infections, this is within 2–3 days after onset of symptoms. Specimens with high concentrations of viral particles, viral antigens, or viral nucleic acid will improve the laboratory's ability to make an accurate diagnosis.

Viruses infecting humans cause a wide range of diseases. Disease signs and symptoms largely influence which specimens are collected for virus testing. Additional factors are epidemiological (travel, exposures, season), and the integrity of the patient's immune system. It is important to recognize that viruses differ in their pathogenic mechanisms, their ability to disseminate, and the organ systems they infect. Specimen collection is guided by the diagnostic test best suited to establish a particular diagnosis. Common sense will often guide collection of the appropriate specimen. For example, if pneumonia is diagnosed, the specimen of choice would be a sputum or lower respiratory tract specimen such as a bronchoalveolar lavage or bronchial brushing. If the patient presents with signs and symptoms of encephalitis, cerebrospinal fluid (CSF) is usually the specimen of choice for viral testing. For patients with suspected viremia, peripheral blood is also routinely submitted. Blood specimens are collected into tubes containing anticoagulant [heparin, acid-citrate-dextrose (ACD) or EDTA]. EDTA is the preferred anticoagulant for obtaining plasma for nucleic acid testing (NAT) since heparin is inhibitory to many nucleic acid amplification chemistries.

Viral infections can be diagnosed by several means. Sometimes, a disease is sufficiently characteristic that an accurate diagnosis can be made from clinical signs and symptoms alone (e.g., chicken pox, measles). However, the clinical presentations of viral infections are often nonspecific. Under these circumstances, laboratory diagnosis is necessary. The laboratory diagnosis may be accomplished by serological, direct detection of antigens or nucleic acid sequences in specimens, and virus isolation in cell culture.

The recommended or commonly used methods to diagnose viral infections depend in large part on the patterns of host immunologic and viral parameters during and after infection. If the amount of infectious virus within a given specimen is very low or absent by the time symptoms appear, culture should not be the primary diagnostic method. If antibody titers are inconsistent due to host factors, prior immunization, latent/recurrent infection, or reinfection, then serology may not be the primary diagnostic method. Another important factor influencing the test method is the ability of a virus to grow in commonly used cell lines using relatively simple cultivation procedures. A number of viruses infecting humans will not grow in standard cell lines (e.g., rotavirus, hepatitis C virus [HCV], parvovirus B19) or require special procedures not available to many laboratories (e.g., Epstein-Barr virus [EBV], human immunodeficiency virus [HIV], rubella virus). For these viruses, non-culture methods such as serology, and antigen or nucleic acid detection are routinely used.

Cell Culture

Historically, the reference method for diagnosis of most viral infections has been culture (isolation). When proper cell lines are used (also referred to as susceptible or permissive cell lines), culture provides highly sensitive detection of many clinically relevant viruses. The main advantage of traditional cell culture is the capacity to isolate a wide variety of viruses using a range of permissive cell lines. Culture is considered nearly 100% specific although cross-contamination of patient specimens or control strains and patient specimens can occur if careful attention is not given to aseptic technique. A positive culture indicates the presence of infectious virus, while non-culture methods (discussed below) may yield positive results in the absence of infectious virus (e.g., detection of viral DNA or RNA). Culture can be important to provide viral isolates for antiviral susceptibility testing, strain typing, and other procedures, both clinical and research.

When a general virus culture is requested, a battery of cell lines must be inoculated because different viruses have various cell replication tropisms. This requires the laboratory to use several different cell lines to provide broad coverage for commonly encountered human viruses. The number and type of cultured cells utilized will depend on the potential viruses that may be found to cause infection in a particular specimen source. The choice of cells is an important aspect of viral culture, as some viruses have very restricted host cell ranges (e.g., CMV), whereas other viruses will replicate on a number of different cell lines (e.g., HSV). Cell cultures are derived from a variety of animal (e.g., human, monkey, rabbit) and organ (e.g., kidney, lung) sources.

Traditionally, clinical specimens were inoculated into capped glass tubes containing a monolayer of cells on one side of the vessel. The cells would be overlaid with cell culture media to maintain cell viability and incubated at 35–37 °C. Developments in cell culture techniques lead to another format for virus isolation using a shell vial. This vessel is a 15 × 45 mm, flat-bottom vial containing a round coverslip on the bottom of the vial with a cell monolayer on the surface. After inoculating vials with patient specimens, they are centrifuged to enhance the entry of virus into susceptible cells by temporarily altering the target cell membrane.

The replication of viruses in cell culture can be detected microscopically by one of several methods. The most familiar is the development of cytopathic effect (CPE). As the name implies, CPE refers to a diseased or abnormal appearance of cells. Viruses often produce characteristic CPE, and the trained virologist is often able to identify viruses based on the appearance of CPE. However, it is recommended (and routine in many laboratories) to confirm the virus using specific methods such as immunofluorescent (IF) antibody staining of scraped and fixed infected cells. The appearance of CPE depends on the type and age of cell line, and the stage of CPE (early vs. late). In addition to morphological changes in the cell monolayer, the time to appearance of CPE and the rate of progression are factors that the experienced technologist will use to help identify a viral isolate. Examples of CPE can be seen in Fig. 2.1. IF staining of cell cultures showing CPE has become a common

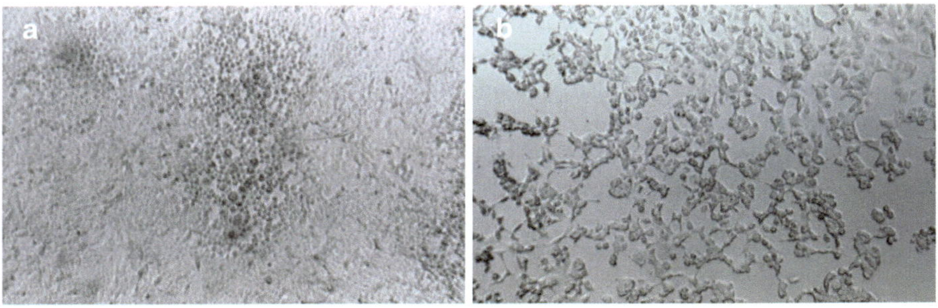

Fig. 2.1 Examples of viral cell culture CPE. The early cell rounding observed with VZV infection evolves into a degenerated cellular appearance that remains as a focal lesion within the cell monolayer (**a**). Adenovirus-infected cells aggregate late in infection into rounded cell clumps, producing grape-like clusters or a lattice-type arrangement (**b**)

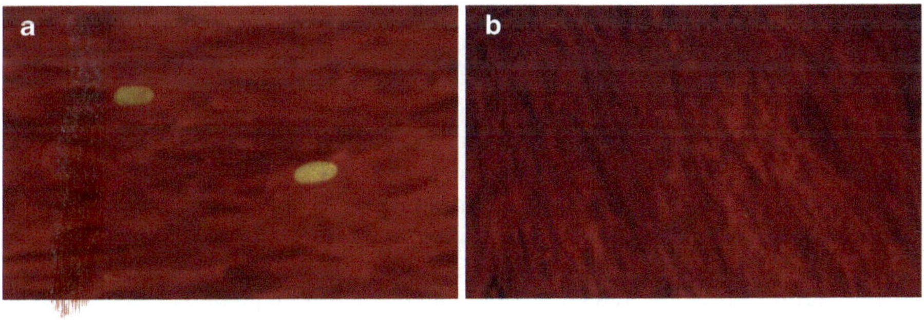

Fig. 2.2 Immunofluorescent staining of a shell-vial cell monolayer with mAbs specific for CMV early antigen in infected cells (**a**) and uninfected cells (**b**)

procedure with the availability of commercially prepared reagents. In this procedure, cells are applied to a slide, and then reacted with a specific monoclonal antibody (mAb) targeted to a viral protein. The monoclonal antibody may be fluorochrome-labeled and the preparation can be examined using a fluorescent microscope.

For rapid detection of viruses, cell cultures can be stained with IF reagents specific for a particular viral antigen prior to the development of CPE. This technique employed with shell vials can significantly shorten the time to detect clinically relevant viruses compared to standard tube cultures from 1 week or longer to 2–3 days. The impact of this format has been important for viruses such as CMV where tube cultures often require incubation for several weeks prior to the development of CPE. Pre-CPE IF staining of shell vials using a fluorescently labeled mAb specific for an early antigen of CMV shows apple-green fluorescence of the nucleus of the infected cell (Fig. 2.2a) compared to an uninfected cell monolayer (Fig. 2.2b).

In some situations, it may be possible to combine or co-cultivate different cell lines within a single vessel to reduce the number of individual cell line vessels required in the laboratory. For example, a co-cultivated cell line used to support the detection of HSV and VZV is comprised of a mixture of African green monkey kidney cells and MRC-5 cells. These shell vials can be examined for CPE up to at least 5 days or, alternatively, can be stained before the appearance of CPE on earlier days to shorten the time to detection. The ability to detect both HSV and VZV in a single culture system is advantageous since the clinical appearance of HSV and VZV skin lesions are often similar.

Antigen Detection

Because of their rapid turnaround time, methods that detect viral antigens directly in patient specimens have become an integral part of the diagnostic virology test regimen. These methods include enzyme immunoassay (EIA) formats and direct specimen testing by IF. The time-to-results for these assays ranges from approximately 15–20 min for EIA in the cartridge or strip format (Fig. 2.3) to 1–2 h for microwell plate EIA and IF assays. This rapid turnaround time allows for point-of-care (POC) use in patient triage and provides results that greatly impact patient treatment and isolation. Unfortunately, in many cases the rapid turnaround time comes at a price- decreased sensitivity or specificity relative to the slower virus isolation methods. Depending on the virus and the specimen source, rapid methods (particularly the EIA format) can have sensitivities as low as 50–80% relative to culture. This is generally the case with rapid antigen detection methods for influenza A and B viruses. In some instances, antigen detection methods are very sensitive, relative to reference cell culture methods. Direct antigen detection methods

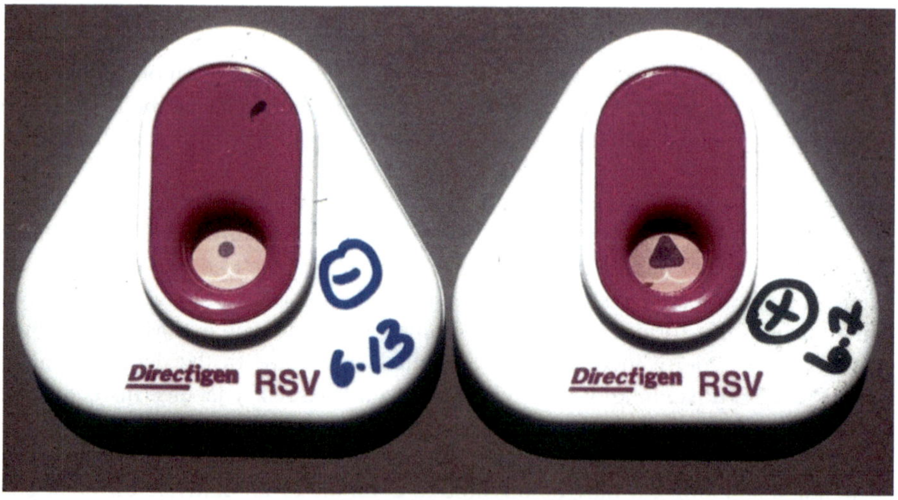

Fig. 2.3 Rapid immunoassay for RSV antigen detection. The patient respiratory specimen is applied directly to the cartridge. After addition of reagents, the colorimetric reaction shows a small circle for negative samples and a triangle for positive ones

such as IF for respiratory syncytial virus (RSV) are generally as sensitive as culture because RSV is a very labile virus and difficult to isolate in cell culture. IF assay is commonly used to detect HSV and varicella zoster virus (VZV) antigens in vesicular skin lesions, and, for VZV, provides sensitivity equivalent to that of virus isolation. For non-cultivatable viruses, antigen detection methods may be an alternative diagnostic option (e.g., rotavirus).

Nucleic Acid Detection

Methods that detect viral nucleic acid sequences in clinical specimens are an important part of the testing regimen of many virology laboratories. Viruses from which nucleic acids are now commonly detected and/or quantified include HSV, enteroviruses, CMV, HIV, HCV, human papilloma virus (HPV), hepatitis B virus (HBV), respiratory viruses, and others. A number of nucleic acid amplification testing (NAT) techniques have been described and commercialized and generally have high analytical and clinical sensitivity, typically greater than that of culture and other non-culture detection methods. The specificity of NAT is also better than antigen detection methods. While the chemistries of these nucleic acid amplification methods differ greatly, the general assay steps are similar: specimen processing to release nucleic acids; amplification of target or probe nucleic acid, or of signal; and detection. However, the performance of molecular assays can vary significantly due to several factors including nucleic acid extraction methods, primer and probe design, amplification and detection technologies, instrumentation, and technical expertise.

The most recognized nucleic acid amplification method is polymerase chain reaction (PCR). The basis of this method is primer-directed replication of DNA. A pair of primers (short oligonucleotides, usually 20–30 bases in length) is designed so that one primer binds to a complementary section of one strand of a double-stranded DNA molecule, while the other binds to the other DNA strand at a distance of typically 100–300 nucleotides apart. In this way, the primers flank a short portion of the viral genome and define the region to be amplified. Each primer serves as the point at which new DNA synthesis begins, using a DNA polymerase enzyme. The PCR mixture also contains deoxyribonucleoside triphosphates (dNTPs), the "building blocks" for DNA, and other components so that a new strand of DNA can be synthesized. PCR requires a series of temperature cycles (thermal cycling), each consisting of a high (~95 °C) temperature and one or two lower temperatures (~50–72 °C). Each temperature cycle completes one DNA replication cycle, in which double-stranded DNA target is melted apart, primers anneal to complementary sequences of each DNA strand, and new DNA strands are synthesized (i.e., primers are extended). The end result is the doubling of the number of DNA copies. PCR results in exponential amplification of target; each round of thermal cycling potentially doubles the amount of the target (amplicon) nucleic acid, so that after 30 cycles as many as a billion copies of newly amplified DNA may exist from a single starting piece of viral genome. The amplified DNA is then readily detected using a labeled DNA probe, visualization in agarose gels, or by other methods. Some techniques utilize fluorescent probes that bind to the amplified DNA region to monitor amplification during each cycle; this is referred to as real-time PCR since amplified DNA is detected in real time (Fig. 2.4). Since many viruses possess RNA genomes, it's necessary to first reverse transcribe the sequence nucleic acid into complementary DNA (cDNA) prior to PCR using a reverse transcriptase (RT) enzyme. Amplification of RNA viruses is referred to as RT-PCR.

Serology

A number of viral diseases are routinely diagnosed by detecting the host antibody response to the pathogen, rather than the agent itself. This method is often used when other techniques that detect the virus or its components are insensitive, the virus is not cultivatable in routine cell culture systems, or culture of the virus is too dangerous. Viral infections detected primarily or at least initially by serologic methods include arboviruses (e.g., West Nile virus [WNV]), hepatitis viruses, HIV, measles (rubeola virus), mumps, and rubella virus). Serologic tests are available in several formats including cartridge- or strip-based immunoassays, microwell plate-based enzyme immunoassays (EIAs), and automated instrumentation for high volume testing.

In a primary infection (i.e., the first time an individual is exposed to a particular virus) antibodies of the IgM class are usually produced first, early in the acute stage of infection, and are the first detectable (Fig. 2.5). IgM antibodies are generally produced for a limited time, and as a result their presence is indicative of recent or current infection. There are exceptions, for example, the IgM response to West Nile virus infection can last a year or longer. IgG antibodies are generally produced after IgM and are often not detectable during the acute stage. Another limitation of IgM assays is that this antibody class is generally produced only in primary (first time)

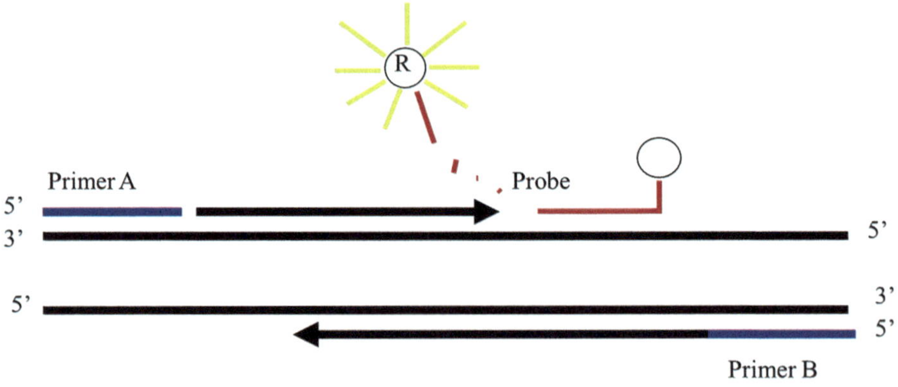

Fig. 2.4 Real-time PCR using hydrolysis probe chemistry

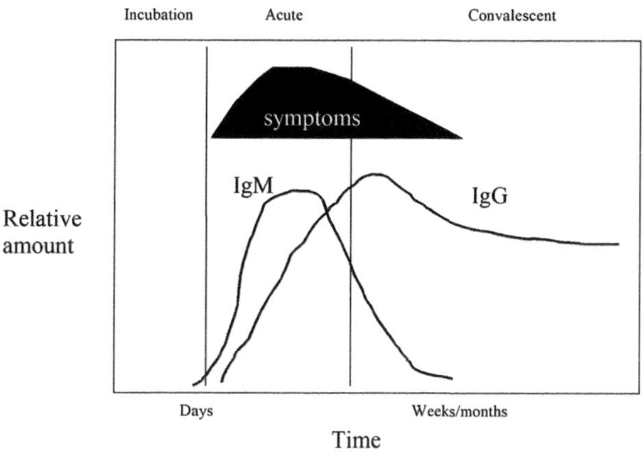

Fig. 2.5 Clinical and antibody response during infection

infections. Reinfection and recurrent infection (such as with latent herpes virus infections) will occasionally elicit an IgM response. IgG antibodies are generally produced after IgM and are often not detectable during the acute stage. Once present, IgG levels often remain detectable for long periods (months to years). Therefore, the mere presence of IgG antibodies in a single specimen cannot distinguish between recent and past infection. However, the IgG antibody titer can sometimes accurately provide this distinction, as extremely high titers can be indicative of recent infection. Antibody titers are measured by preparing serial two-fold dilutions (1:2, 1:4, 1:8, 1:16, etc.) and testing each by a particular serologic assay. The titer is the greatest dilution that is positive (reactive) by the assay. Generally, the most specific serologic diagnosis of infection is provided by measuring IgG titers on paired specimens; one collected during the acute stage and another 2–4 weeks later during convalescence. Seroconversion (i.e., IgG negative in the acute sample and IgG positive in the convalescent sample), or a four-fold or greater increase in the IgG titer in these specimens is considered to be conclusive evidence of infection.

Antibodies can be detected and quantified by a number of methods, including EIA, immunofluorescence (IF), complement fixation (CF), chemiluminescent immunoassay (CIA), and neutralization. Because of important assay characteristics like speed, throughput, simplicity, accuracy, cost, and availability of commercial kits, EIA and CIA are the most commonly used methods for diagnosing or screening for infection. Similar to antigen detection tests, serologic tests are performed using microwell plates or membranes as solid substrates (EIA or CIA) or using slides as solid substrates (IF). The substrates are provided pre-coated with viral antigen. The antigen can be crude viral lysate, a purified single viral antigen, or recombinant antigen. Patient serum (single dilution for screening or multiple serial dilutions to determine titer) is added to the microwell, membrane cartridge, or slide; and patient antibody will bind to the antigen if it is recognized. A washing step will eliminate unbound antibody. In the next step, a fluorochrome or enzyme-labeled, anti-human antibody is added (direct format). The anti-human antibody binds to any antibody already bound to antigen, and the complex is detected by adding substrate and detecting color generation (EIA), or viewing under a fluorescent microscope (IF).

The plaque reduction neutralization test (PRNT) is the most common type of neutralizing assay that involves inoculating a serially diluted sample of live virus in a monolayer of virus-susceptible cells. A lack of plaque formation by the live virus indicates the presence of neutralizing antibodies. Although labor intensive, this is a specific method for testing for viral infection and can be used as a confirmatory test in presumptive positive cases by EIA. PRNTs are used as confirmatory tests for flaviviruses such as Zika virus, Dengue virus, and WNV due to the potential for cross-reactivity among viruses in this genus. Because testing is labor intensive and often requires at least biosafety level (BSL) 3 laboratory containment, testing is limited to reference and public health laboratories.

Antiviral Susceptibility Testing

Antiviral susceptibility testing may be needed when patients undergoing antiviral therapy continue to deteriorate or the concentration of virus causing infection continues to climb. In certain situations, this can be associated with drug-resistant virus. Antiviral resistance can be defined as a decrease in susceptibility to a particular drug that can be determined by *in vitro* testing. This may be accomplished by employing phenotypic or genotypic testing methods. Phenotypic assays are those that measure the inhibitory effect of antiviral agents on the entire population in a patient isolate. Genotypic assays analyze viral nucleic acid to detect specific mutations that cause antiviral drug resistance.

Phenotypic Methods

The plaque reduction assay (PRA) is the classic phenotypic method to assess inhibition of viral isolates in the presence of antiviral agent. The PRA utilizes viral-infected cell cultures to which varying concentrations of the antiviral agent is added. After an appropriate period of incubation, the extent of viral replication is measured by determining the concentration of antiviral agent inhibiting plaque formation by 50%. This concentration of drug is considered the IC50 value (also known as 50% inhibitory concentration or 50% effective concentration).

The recombinant virus assay (RVA) has been used predominantly to assess the susceptibility of HIV to nucleoside reverse transcriptase inhibitors (NRTIs) and protease inhibitors (PIs). The RVA is a phenotypic assay that involves RT-PCR amplification of the complete HIV reverse transcriptase (RT) and protease (PR) gene coding sequences directly from the patient's plasma. Typically, a viral load of at least 500–1000 copies of RNA/mL of plasma is required. The RT and PR genes are then ligated into an HIV viral vector and co-transfected with a plasmid containing the envelope proteins of another retrovirus (murine leukemia virus) into a permissive cell line. Thus, the virus generated (a chimeric virus) will contain the patient's HIV strain RT and PR gene sequences in a background of an HIV strain from which the original RT and PR sequences have been deleted. The ability of this chimeric virus to replicate in the presence of various levels of NRTIs and PIs is measured and expressed as an IC50 value. The RVA is labor intensive and requires approximately 2 weeks to complete.

Genotypic Methods

CMV: Ganciclovir resistance in CMV is associated with mutations in the viral genes encoding UL97 (phosphotransferase) and UL54 (DNA polymerase). UL54 mutations also encode resistance to cidofovir and foscarnet. Both genes of a patient's CMV strain can be sequenced, either directly from a clinical sample if there is enough DNA or a viral culture isolate, and compared to a database of known mutations conferring resistance.

HIV: Resistance to NRTIs and PIs is associated with mutations in the HIV RT and PR genes, respectively. One approach to map these mutations involves directly sequencing the RT and PR genes (approximately 1.5 kb) of a patient's HIV strain and comparing them to a library of known resistance mutations. Another approach compares the RT and PR sequences to a large database of known genotypes and phenotypes to predict a phenotype based on sequence data (i.e., a virtual phenotype). A concern with either approach involves the inability to detect mutant variants present at low frequencies. In addition, genotypic assays can detect only known resistance-associated mutations and the complex interactions among different combinations of mutations may not allow for accurate determination of resistance.

Influenza Viruses: Both conventional sequencing and pyrosequencing have been used to identify mutations associated with resistance to antivirals used to treat influenza virus infections. Oseltamivir resistance in seasonal and 2009 pandemic influenza A H1N1 strains is most commonly associated with the H275Y mutation in the viral neuraminidase gene. Other neuraminidase gene mutations may confer oseltamivir resistance in other subtypes of influenza A virus and influenza B virus.

Summary

Since no single approach is optimal for detecting all viruses in all clinical situations, it's often important to combine multiple methodologies to optimize viral disease diagnosis, yielding medically useful and cost-effective viral testing results. Culture is generally more time- and labor-intensive than other methods but still plays a role in some situations such as providing clinical isolates for epidemiologic studies, for the establishment of vaccine candidates, to evaluate mechanisms of antiviral resistance for new antiviral drugs, and to identify potential novel viral agents. Over the last decade, NAT has become more widely available in diagnostic laboratories and can provide more timely, sensitive, and highly specific results. This information is important since for many viral infections testing should be completed within a time frame that can affect patient management (i.e., initiation or discontinuation of antivirals), provide for efficient hospital bed utilization and appropriate infection control measures to reduce nosocomial transmission, and identify outbreak situations.

Case 1

Marc Roger Couturier

A previously healthy and active 53-year-old male presented to the local emergency department in September with a chief complaint of 1 week of progressively worsening fatigue, nausea, dizziness, and fever. He developed visual changes (including photophobia and blurriness) on the day of admission and was admitted to rule out meningitis and for evaluation of the source of fever. His only significant recent medical history was Graves' disease which was treated with radioactive iodine ablation 2 years prior, for which he was not prescribed thyroid replacement hormone. The patient was seen 2 days prior to admission at an ambulatory care clinic and was prescribed levothyroxine based on previously unmanaged thyroid disease. At admission the patient had a 40.1 °C fever, chills, difficulty breathing, a somewhat stiff neck, and sore throat. The patient noted nausea and constipation for 7 days but denied abdominal pain, vomiting, or diarrhea. Initial laboratory results were mostly normal, with elevated TSH, creatinine, AST, protein, albumin, and alkaline phosphatase (Table 2.1). The patient was tested by rapid molecular tests for influenza A & B by nasal swab and group A streptococcus by pharyngeal swab, which were negative. Blood cultures were drawn at admission and remained negative through the course of his hospitalization. There was a low suspicion for bacterial meningitis, so a lumbar puncture was not obtained at admission. Head and chest CTs were unremarkable. An acute hepatitis antibody panel was ordered due to the elevated AST and alkaline phosphatase and was negative for hepatitis A, hepatitis B, and hepatitis C viruses.

On hospital day 2, the patient developed labored breathing and became dizzy and confused. He continued to deteriorate throughout the day and mild cerebral hemorrhage was seen on head CT with contrast. On hospital day 3, he suffered further encephalopathy resulting in respiratory arrest following a lumbar puncture and pulseless electrical activity. The CSF obtained on day 3 was consistent with a viral meningitis: clear in appearance with a normal opening pressure (120 mmHg), elevated WBC (915 cells/μL, 53% lymphocytes), elevated total protein (90 mg/dL), and normal glucose (67 mg/dL). The following tests were ordered on the CSF specimen: VZV and HSV1/2 PCR, meningitis/encephalitis PCR panel, West Nile virus (WNV) PCR, and WNV IgM ELISA. Subsequent MRI imaging was consistent with diffuse hypoxic brain injury and the family elected for comfort care measures only on hospital day 4. The patient passed away shortly thereafter. On hospital day 4, the

Table 2.1 Initial blood laboratory results at the time of admission

Result	Value	Ref range
Sodium	137	136–144 mmol/L
Potassium	3.7	3.3–5.0 mmol/L
Chloride	102	102–110 mmol/L
Carbon dioxide	24	20–26 mmol/L
BUN	11	8–24 mg/dL
Creatinine	**1.47**[a]	0.72–1.25 mg/dL
Glucose	108	64–128 mg/dL
Calcium	9.7	8.4–10.5 mg/dL
Protein (total)	**9.3**[a]	6.5–8.4 g/dL
Albumin	**5.2**[a]	3.5–5.0 g/dL
Bilirubin (total)	1.1	0.2–1.4 mg/dL
Alkaline phosphatase	**127**[a]	38–126 U/L
AST	**118**[a]	16–40 U/L
ALT	57	5–60 U/L
Anion gap	11	8–14 mmol/L
TSH	**13.83**[a]	0.35–4.94 mU/L

BUN blood urea nitrogen, *AST* aspartate aminotransferase, *ALT* alanine transaminase, *TSH* thyroid-stimulating hormone

[a] Indicates abnormal result

results from all CSF tests returned from the laboratory (Table 2.2). The WNV IgM in CSF was weakly positive (1.88, cutoff 1.1). A presumptive cause of death was proposed as West Nile neuroinvasive disease. Confirmatory testing of the CSF at the local state health lab revealed plaque reduction neutralization (PRNT) antibody titers specific for WNV and negative for St. Louis encephalitis virus, confirming the diagnosis.

Discussion

West Nile virus is an enveloped, positive-sense single-stranded RNA virus of the family *Flaviviridae*. Like many other members of the *Flaviviridae*, WNV is transmitted by a mosquito, specifically of the genus *Culex*. The virus cycles normally between *Culex* and avian hosts. Both humans and members of the equine family can serve as susceptible "dead-end" hosts, with each potentially showing disease symptoms but not sustaining sufficient viremia to allow subsequent transmission to uninfected mosquitos. No other mammals are known to serve as susceptible hosts for WNV. Humans may also acquire WNV from transfusions of blood and platelets, as well as through organ transplantation.

Most WNV-infected humans are clinically undocumented due to very mild or asymptomatic infections. Retrospective serological studies have revealed that the majority of individuals with WNV antibodies do not recall a recent febrile illness.

Table 2.2 Infectious disease test results performed from the CSF specimen collected at day 4 of admission

Test method	Target	Result
Herpes PCR	HSV 1	Not detected
	HSV 2	Not detected
Varicella PCR	VZV	Not detected
Multiplex meningitis/encephalitis PCR	EV	Not detected
	hPEV	Not detected
	VZV	Not detected
	CMV	Not detected
	HSV 1	Not detected
	HSV 2	Not detected
	HHV-6	Not detected
	Listeria monocytogenes	Not detected
	Streptococcus agalactiae	Not detected
	Streptococcus pneumoniae	Not detected
	Escherichia coli K12	Not detected
	Neisseria meningitidis	Not detected
	Haemophilus influenzae	Not detected
	Cryptococcus neoformans/gattii	Not detected
West Nile virus PCR	WNV	Not detected
West Nile virus IgM MAC-ELISA	WNV	**Detected**[a]

HSV herpes simplex virus, *VZV* varicella zoster virus, *EV* enterovirus, *hPEV* human parechovirus, *CMV* cytomegalovirus, *HHV-6* human herpesvirus 6
[a] Indicates abnormal result

In more severe presentations such as neuroinvasive disease, many cases are still not initially diagnosed due to provider unfamiliarity with WNV disease and/or the associated/appropriate testing methods. As a result, many cases are documented post-mortem. West Nile fever (WNF) and West Nile neuroinvasive disease (WNND) are the two symptomatic clinical courses of disease which are typically seen as sudden onset syndromes occurring between July and September in northern hemisphere of the Americas. For WNF, the following symptoms may occur in 50% of patients: headache, weakness, morbilliform (macules and papules typically distributed on the trunk) or maculopapular rash, fever, and/or myalgias. These symptoms typically last days to weeks. WNND occurs in less than 1% of infected patients. Patients of the male sex, advanced age, and/or persons having underlying immune compromising conditions are more likely to develop WNND. Meningitis or encephalitis may be seen in WNND with symptoms including altered mental state, lethargy, acute flaccid paralysis, and radiologically confirmed meningeal inflammation. Other accompanying symptoms can include nausea and vomiting, myalgia, arthralgia, fever, and rarely a rash.

The diagnosis of WNV is challenging, and typically is achieved by detecting IgM antibody in serum (WNF and WNND) or CSF (WNND). Due to the high degree of relatedness between WNV and several other arthropod-borne

Flaviviridae, the specificity of IgM tests is often a limitation. Though cross-reactivity can be detected with serum from patients with recent (<1 year from disease) Japanese encephalitis virus and dengue virus infections, these viruses typically do not co-circulate in areas of high prevalence for WNV (e.g., North America, northern Africa, and eastern Europe). Recent vaccination with the Yellow Fever or Japanese encephalitis vaccines may also cause false-positive results. In North America, St. Louis encephalitis virus and deer tick virus (Powassan virus lineage II) pose the greatest risk for nonspecific, cross-reactive antibodies. Commercially available WNV IgM ELISAs typically are IgM antibody capture (MAC) ELISAs to provide a more specific and sensitive detection. Even still, these results must be confirmed by plaque reduction neutralization testing (PRNT) for all possible cocirculating/cross-reactive viruses to definitively determine which flavivirus is causing the illness. In some cases, even PRNT cannot definitively differentiate the causative virus.

PCR for WNV is generally considered of limited clinical value on CSF and serum/plasma/whole blood for any manifestation of disease. In general, PCR is most useful in the first 2–5 days of illness while the viral replication is highest. The exception is in patients who have immunosuppression, for whom the viremic window may be much longer (up to 10–14 days). In these patients, it is common to have delayed or diminished production of IgM antibody. In otherwise healthy patients with WNF, PCR has very little value. In fact, in these patients, by the time they seek care, the serum IgM should be detectable. IgM detection only suffers from low sensitivity if the patient it tested too early in the disease course (generally before 7 days of symptom duration). In these cases, PCR may have better sensitivity. Importantly, no FDA-cleared WNV PCR tests exist for patient testing, and commercially available CSF multiplex PCR assays do not include WNV as a target. However, in the United States there are approved nucleic acid tests for WNV to routinely screen blood donor products to prevent transfusion-associated transmission. With the increasing reliance on such rapid multiplex panels that target common CSF pathogenic viruses (e.g., enterovirus, HSV, VZV, HHV-6), physicians must remember the possibility of WNV and the need to order IgM tests (or PCR in limited settings) if the disease is clinically compatible and the exposure history supports this virus in the differential diagnosis.

Summary Points

- West Nile virus is an arthropod-borne virus that is transmitted most frequently during the late summer months and is more severe in elderly and immunocompromised patients.
- IgM antibody detection by MAC-ELISA with confirmation by PRNT is the most reliable method of diagnosis.
- WNND is best diagnosed by CSF detection of IgM whereas WNF is diagnosed by serum testing for IgM.

- PCR has a limited scope of utility for diagnosis of WNF and WNND for CSF or serum/plasma/whole blood. Clinical sensitivity is limited by a short viremic phase in otherwise healthy patients.

Selected References

Curren EJ, Venkat H, Sunenshine R, Fitzpatrick K, Kosoy O, Krow-Lucal E, Adams L, Kretschmer M, Fischer M, Hills SL. Assessment of immunoglobulin M enzyme-linked immunosorbent assay ratios to identify West Nile Virus and St. Louis Encephalitis Virus infections during concurrent outbreaks of West Nile Virus and St. Louis Encephalitis Virus diseases, Arizona 2015. Vector Borne Zoonotic Dis. 2020;20(8):619–23.

Endale A, Medhin G, Darfiro K, Kebede N, Legesse M. Magnitude of antibody cross-reactivity in medically important mosquito-borne flaviviruses: a systematic review. Infect Drug Resist. 2021;14:4291–9.

Peterson LR, Brault AC, Nasci RS. West Nile Virus: review of the literature. JAMA. 2013;310(3):308–15.

Piantadosi A, Kanjilal S. Diagnostic approach for arboviral infections in the United States. J Clin Microbiol. 2020;58(12):e01926-19.

Case 2

James J. Dunn

A 17-year-old male presented to the emergency room with a 4 day history of fever, lower back and abdominal pain, and a 1 day history of nausea and vomiting that appeared to have streaks of blood. He had a past medical history that included asthma and environmental and food allergies, and three previous surgeries for recurrent sinus infections. Given the current episodes of hematemesis (blood in vomit), there was concern for possible gastrointestinal bleeding or a gastric ulcer. Laboratory tests on admission showed normal cell counts (WBC and RBC) and normal values for the common chemistry tests. Of note, the C-reactive protein (CRP) level was elevated at 4.0 mg/dL (reference range, <1.0 mg/dL).

Shortly after admission an upper endoscopic procedure was performed. Tissue biopsies were obtained from areas of the proximal and distal esophagus and the gastric cavity. Images taken with the endoscope during the procedure revealed

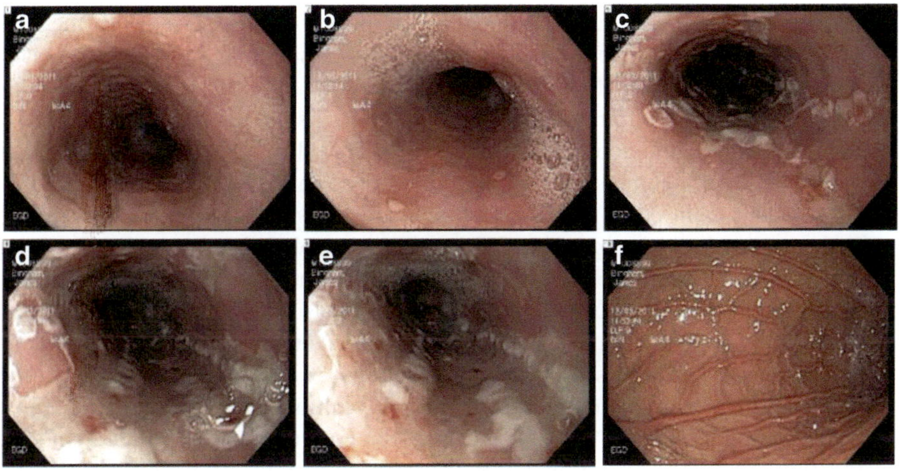

Fig. 2.6 Upper endoscopy images showing discrete oval-shaped mucosal erosions with whitish necrotic bases in the proximal esophagus. The severity and frequency of lesions increase down to the mid-esophagus (**a–c**), and the distal esophagus has complete mucosal erosion and mucosal edema with more intense linear erosions (**d**, **e**). The stomach was not affected (**f**)

discrete oval-shaped mucosal erosion with whitish necrotic bases. The severity and frequency of lesions increased down to the mid-esophagus and the distal region had complete mucosal erosion and edema with more intense linear erosions (Fig. 2.6). Histologic examination of the tissue biopsies obtained during the procedure revealed ulcerative exudative esophagitis (Fig. 2.7a) and large, multinucleated epithelial cells with ground glass intranuclear inclusion bodies consistent with cytopathic effect due to herpes simplex virus (HSV) (Fig. 2.7b). HSV type 1 (HSV-1) was detected using PCR and viral culture methods on the biopsy specimens. The patient was subsequently started on intravenous acyclovir and showed improvement of symptoms within 24 h. Serologic testing for HSV antibodies later revealed that the patient had high levels of both IgM and IgG against the virus.

Discussion

HSV is an enveloped virus whose genome consists of linear double-stranded DNA. The virus consists of two antigenically distinct types, HSV-1 and HSV-2. HSV primarily infects epithelial cells of mucosal membranes, but can become systemic infecting internal organ systems. While both types can infect any mucosal surface (oral or genital), HSV-1 infections are primarily associated with the oral mucosa (fever blisters or cold sores), whereas HSV-2 is more commonly associated with urogenital mucosa. HSV-1 and HSV-2 are both commonly associated with genital infections which are considered sexually transmitted (herpes). The histopathologic changes in mucosal lesions caused by HSV-1 or HSV-2 replication result in cytoplasmic enlargement, nuclear alterations and may lead to the formation of multinucleated giant cells. The local inflammatory response is frequently neutrophilic in dermatologic specimens and induces edema and vesicle formation. Infected cells often contain intranuclear ground glass viral inclusions (Fig. 2.7a), but intranuclear Cowdry type A inclusions (i.e., eosinophilic structure surrounded by a clear area imparting a targetoid appearance) may also be seen. Lesions may also demonstrate the classic three M's of viral cytopathic effect due to HSV which include nuclear molding, margination of chromatin, and multinucleation (Fig. 2.7b). Since the viral cytopathic effects of HSV and varicella zoster virus (VZV) are identical, histopathology should be correlated with ancillary laboratory testing (e.g., PCR) and clinical symptomatology.

The virus is spread by direct contact with oral or genital secretions that contain infectious virions. HSV can be transmitted through secretions even if lesions are not present. Most infections with HSV do not result in overt clinical manifestations, either at the time of primary acquisition or during reactivation. Primary infection with HSV-1 is commonly acquired in early childhood with worldwide prevalence estimated to be greater than 50%. When symptoms are present, primary infection typically presents with oral lesions that progress to ulcerations and heal within 2–3 weeks. The prevalence of HSV-2 is more variable since transmission is associated with sexual activity. After primary infection, the virus develops latency in the sensory ganglia neurons: the trigeminal ganglia for oral infections and lumbosacral dorsal root ganglia for urogenital infections. Once latency is established, episomal viral DNA is maintained within the neuronal cells for the lifetime of the host. Reactivation of HSV infection is associated with illness, stress, fatigue, immunosuppression, menstruation, or trauma. During reactivation, the virus travels down the neurons to mucocutaneous sites and when clinically manifested leads to a single or multiple vesicles, close to the site of primary infection. In immunocompetent individuals, reactivation is typically asymptomatic or presents with milder symptoms and more rapid healing of the lesions than with primary infection. Invasive HSV disease can lead to central nervous system infection or systemic illness with multi-organ involvement.

HSV esophagitis is the second most common cause of infectious esophagitis behind *Candida* yeast infection. It occurs more commonly in immunocompromised hosts such as those with HIV/AIDS, organ transplant or cancer and is associated with reactivation of the virus following immunosuppressive therapy. HSV esophagitis can be life-threatening in immunocompromised patients, so it is important to make a rapid diagnosis so that appropriate therapy can be initiated. In immunocompetent hosts, HSV esophagitis is quite rare but most cases are seen

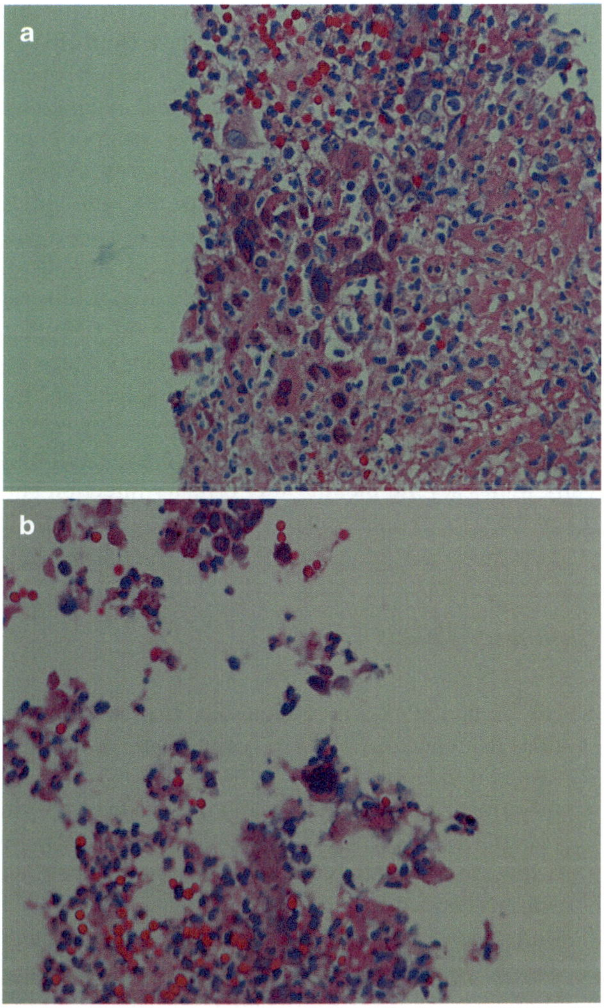

Fig. 2.7 Distal esophagus histopathology showing an inflammatory exudate with many viral-infected cells with intranuclear ground glass inclusions (**a**) (H&E, magnification 400×) and the classic three M's of HSV cytopathic change including molding, margination and multinucleation (**b**) (H&E, magnification 1000×)

in adolescents and adults. In these cases, it is almost always a primary infection with HSV-1 causing esophageal infection and IgM seropositivity can often be demonstrated. HSV esophagitis can be a self-limiting disease in otherwise healthy, immune competent individuals but may require 2–3 weeks for full recovery. Antivirals such as acyclovir can be administered in severe cases to reduce the duration of symptoms. Acyclovir is a nucleoside analog that when activated by the viral thymidine kinase enzyme inhibits replication of HSV by preventing viral DNA synthesis.

Historically, HSV has been cultured from clinical specimens using traditional viral cell culture techniques or shell vials. Both HSV-1 and HSV-2 readily and rapidly (within 1 or 2 days) replicate in a variety of cell lines including A549 and human fibroblasts. The characteristic focal cytopathic effects (CPE) in viral culture include disruption of the cell monolayer with rounding and enlargement of infected cells. Definitive identification of HSV in cell culture requires immunofluorescent staining with HSV-1- and HSV-2-specific antibodies. Fluorescent antibody test reagents targeting HSV-specific antigens directly in vesicular skin lesion material is also commercially available. Nucleic acid amplification tests like PCR are now commonly available and more routinely used for the rapid, direct detection, and differentiation of HSV-1 and HSV-2 from blood, CSF, tissue, and swabs from vesicular lesions. PCR is much more sensitive than culture or direct immunofluorescence testing for detection of HSV and is considered the "gold standard" for certain syndromes such HSV encephalitis. Serology testing is available, but epitopes are highly conserved between HSV-1 and HSV-2 often resulting in cross-reactivity. IgM typically wanes several months after primary infection but may reappear in some cases when the virus is reactivated, while IgG titers are maintained for years.

Summary Points

- HSV-1 and HSV-2 are ubiquitous viruses transmitted via oral or genital secretions from persons who are symptomatic or asymptomatic with primary or reactivated infections.
- Infections with HSV-1 typically involve mucosal sites around the face and upper extremities whereas HSV-2 infections generally involve the genitalia and skin below the waist, but both types can be associated with sexually transmitted infections (STIs).
- Both HSV-1 and HSV-2 can cause mucocutaneous infections as well as invasive diseases such as meningitis and encephalitis.
- HSV esophagitis is a rare entity, particularly in immunocompetent individuals, but can be life-threatening because of difficulties swallowing.
- Nucleic acid amplification tests such as PCR have become the mainstay for diagnosis of HSV infections using a variety of clinical specimen types such as CSF, blood, tissue, and mucocutaneous vesicular fluid.

Selected References

Anderson NW, Buchan BW, Ledeboer NA. Light microscopy, culture, molecular, and serologic methods for detection of herpes simplex virus. J Clin Microbiol. 2014;52:2–8.

Canalejo E, Garcia Duran F, Cabello N, Garcia Martinez J. Herpes esophagitis in healthy adults and adolescents: report of 3 cases and review of the literature. Medicine (Baltimore). 2010;89:204–10.
Fatahzadeh M, Schwartz RA. Human herpes simplex virus infections: epidemiology, pathogenesis, symptomatology, diagnosis, and management. J Am Acad Dermatol. 2007;57:737–63.
Rmanathan J, Rammouni M, Baran J, Khatib R. Herpes simplex virus esophagitis in the immunocompetent host: an overview. Am J Gastroenerol. 2000;95:2171–6.

Case 3

Angela Ma and Marc Roger Couturier

A 4-month-old male presented to the pediatric emergency department with symptoms of acute gastroenteritis. His parents described the patient having experienced 3 days of vomiting, mild, non-bloody diarrhea (three stool passages within the last 24 h), and failure to thrive. Upon physical examination, the patient was found to be hypertensive with a blood pressure of 134/90 mmHg and a fever of 38.4 °C. Notable dehydration was observed by the care team, for which intravenous fluids and ondansetron were provided to prevent further nausea and vomiting. A gastrointestinal pathogens PCR panel was ordered on a stool specimen in addition to a complete blood count (CBC) with differential, chemistry panel, and urinalysis, of which the latter two revealed no abnormalities. From the CBC with differential results, the patient's total lymphocyte count was markedly decreased at 2750 cells/μL (reference range: 5500–17,000 cells/μL), which prompted additional immunological tests including T-cell and B-cell immunodeficiency profile testing and quantitative immunoglobulins, to be ordered. Accordingly, the patient was admitted, and infectious disease and immunology consults were requested.

From the initial examination and discussion with his parents, the patient was primarily breastfed by his mother, with occasional supplementation with formula. His parents denied starting the patient on solid foods at this point. After further inquiry into the patient's medical history, it was discovered that the patient's family had recently moved to the United States from India, when the patient was 8 weeks old. The extent of pre- and postnatal care the mother and patient received in India could not be determined in depth without the aid of an interpreter, but the parents did remark the patient was born without complications. The parents also mentioned that the patient developed recurrent upper and lower respiratory infections but attributed them to exposures early in life while living in India. Since arriving in the United States, the patient had been receiving vaccinations according to the US child and adolescent immunization schedule and had recently received his second dose of the RotaTeq rotavirus vaccine 2 weeks ago.

Results of the gastrointestinal pathogens PCR panel returned from the laboratory after 4 h and was positive for rotavirus A (Table 2.3). After receiving the results, the

Table 2.3 Gastrointestinal pathogens PCR panel results from stool specimen collected upon admission

Classification	Organism	Result
Bacteria	*Campylobacter* (*C. jejuni*/*C. coli*/*C. upsaliensis*)	Not detected
	C. difficile Toxin A/B	Not detected
	Plesiomonas shigelloides	Not detected
	Salmonella	Not detected
	Vibrio (*V. parahaemolyticus*/*V. vulnificus*/*V. cholerae*)	Not detected
	Vibrio cholerae	Not detected
	Yersinia enterocolitica	Not detected
Diarrheagenic *E. coli*/*Shigella*	Enteroaggregative *E. coli* (EAEC)	Not detected
	Enteropathogenic *E. coli* (EPEC)	Not detected
	Enterotoxigenic *E. coli* (ETEC)	Not detected
	Shiga toxin-producing *E. coli* (STEC)	Not detected
	E. coli O157	Not detected
	Shigella/Enteroinvasive *E. coli*	Not detected
Parasites	*Cryptosporidium*	Not detected
	Cyclospora cayetanensis	Not detected
	Entamoeba histolytica	Not detected
	Giardia lamblia	Not detected
Viruses	Adenovirus 40/41	Not detected
	Astrovirus	Not detected
	Norovirus GI/GII	Not detected
	Rotavirus A	**Detected**[a]
	Sapovirus	Not detected

[a] Indicates abnormal result

care team ordered the rotavirus genotyping test performed at the Centers for Disease Control and Prevention (CDC) that confirmed vaccine-derived infection from the pentavalent RotaTeq vaccine. Follow-up testing for the rotavirus antigen by enzyme immunoassay (EIA) remained detectable up to 6 weeks after initial diagnosis, suggesting prolonged shedding of the virus from patient stool. During the patient's hospitalization, immunologic workup showed near absent T lymphocyte and NK cell counts but normal B lymphocyte counts, and decreased IgG and IgA immunoglobulins (Table 2.4), resulting in a preliminary diagnosis of $T^{-}B^{+}NK^{-}$ severe combined immunodeficiency (SCID). Follow-up testing for T-cell receptor excision circle (TREC) was absent and supported by a lack of thymic tissue observed in radiographical imaging. Genetic testing ultimately identified the patient as having SCID associated with the X-linked IL2RG genotype. Intravenous immunoglobulin and prophylaxis (antibacterial, antiviral, antifungal) therapy were initiated to the patient in efforts to prevent and minimize additional infections. The patient later underwent successful haploidentical hematopoietic stem cell transplantation at 8 months of age.

Discussion

Rotavirus is a member of the *Reoviridae* family and was first discovered in the 1950s from rectal swabs collected from monkeys and identified in humans in 1973. The virus is non-enveloped and composed of three concentric shells that enclose a linear, 11-segment double-stranded RNA genome. On the outermost shell are the capsid proteins VP7 and VP4, which are also known as the surface and spike proteins, respectively. These proteins serve as antigens that induce the

Table 2.4 Immunological test results of patient

Test	Patient results	Reference range[a]
Lymphocytes	2750 cells/μL	5500–17,000 cells/μL
T cells ($CD3^+$)	882 cells/μL	2200–9200 cells/μL
Naïve (CD45RA)	6%	66–100%
Memory (CD45RO)	92%	1–42%
T helper cells ($CD4^+$)	352 cells/μL	1600–6500 cells/μL
Cytotoxic ($CD8^+$)	247 cells/μL	300–3400 cells/μL
B cells ($CD19^+$)	1969 cells/μL	520–2300 cells/μL
NK cells	76 cells/μL	97–2000 cells/μL
Immunoglobulin A	4 mg/dL	2–126 mg/dL
Immunoglobulin G	106 mg/dL	242–1108 mg/dL
Immunoglobulin M	70 mg/dL	21–215 mg/dL
TREC	Absent	Present

[a] Reference range of 91–180 days for complete blood count, 2–4 months of age for T- and B-cell immunodeficiency profile testing, 0–2 years of age for quantitative immunoglobulins

production of neutralizing antibodies involved in immune protection. Capsid protein VP6 forms the intermediate capsid layer of rotavirus and is also the most immunogenic protein component of the virus. Although VP6 doesn't induce the formation of neutralizing antibodies, the different antigens expressed on the protein are used as the basis for which rotavirus is classified into seven different groups. The inner capsid layer is comprised of the VP1-3 proteins; VP1 and VP3 are enzymes involved in viral replication of rotavirus, whereas VP2 makes up the inner core of the capsid.

Based on the antigenic properties of the VP6 protein, rotavirus is divided into groups A through G, of which groups A, B, and C infect humans. G1P[8], G2P[4], G3P[8], G4P[8], and G12P[8] genotypes of rotavirus group A are responsible for the majority of infections in North America. Human infection with rotavirus groups B and C are less frequent than group A and are mainly associated with food- and waterborne outbreaks from infected swine. Rotavirus can be transmitted by the oral-fecal route through close contact with infected people, commonly within the same household, or indirectly by contact with contaminated fomites such as shared utensils and surfaces. Nosocomial transmission of rotavirus is a significant concern, and it is estimated that up to 50% of children hospitalized for gastroenteritis are positive for rotavirus. In temperate climates, seasonality is observed with rotavirus infection peaking in fall/winter months whereas in tropical regions rotavirus infection occurs year-round. Replication of the virus occurs in the villous epithelium of the small intestine and patients present with varying degrees of diarrhea, vomiting, and/or fever. Infection is typically mild and self-limiting within 3–7 days. However, prolonged illness in children between 6 months and less than 2 years of age and immunocompromised patients, such as individuals with HIV/AIDS or solid organ/stem cell transplant recipients, can lead to severe dehydrating diarrhea or chronic infection, respectively. Infants less than 3 months of age present with lower rates of rotavirus infection due to passive transfer of maternal antibodies. Rotavirus is an important cause of early childhood nonbacterial gastroenteritis and in developing countries it is a significant cause of infant mortality due to complications associated with decreased intestinal absorption, electrolyte imbalance, metabolic acidosis, and isotonic diarrhea. Treatment for rotavirus gastroenteritis is primarily oral rehydration. Shedding of rotavirus can occur before and after acute illness and due to the virus's hardiness and resistance to inactivation it can survive in the environment for weeks to months. Reinfection is not uncommon, due to the limited immunity that is conferred following recovery from infection. However, subsequent rotavirus infections are typically not as severe.

Two live rotavirus vaccines are available for prevention of childhood infection. Rotarix is an attenuated monovalent vaccine containing genotype G1P[8] that is administered in two doses orally, typically at 2 and 4 months of age. RotaTeq is a pentavalent human bovine vaccine that contains the G1P[5], P[8]G6, G2P[5], G3P[5], and G4P[5] reassortants and consists of three doses orally given at 2, 4, and 6 months of age. In 2006, vaccination with RotaTeq was implemented in the US vaccination program and since then, a decline in laboratory-confirmed rotavirus A infections has been observed by greater than 50%. Both Rotarix and RotaTeq vaccines have been noted to be less effective at preventing rotavirus infection in Africa and Asia compared to North America and Europe due to geological differences in predominant circulating genotypes. Following the implementation of rotavirus A vaccines into immunization programs, vaccine-derived pediatric cases of infection have been confirmed through the identification of G1P[8] strains in patient stool. Infections have been documented in symptomatic, vaccinated infants and on rare occasion in unvaccinated close contacts of recently vaccinated infants. In many of these cases, infected individuals have an undiagnosed primary immunodeficiency. Patients without the ability to mount an immune response to the live-attenuated vaccines consequentially experience symptomatic illness. The most common underlying primary immunodeficiency associated with vaccine-derived rotavirus infection is SCID; however, there have been reports of vaccine-derived infection where no underlying cause can be determined. In 2009 and 2010, manufacturers of both the RotaTeq and Rotarix vaccines revised the prescribing information and patient labeling of the vaccines to include SCID as a contraindication for administration of rotavirus vaccines. Symptomatic patients with vaccine-derived infection present with persistent diarrhea and vomiting but are less likely to progress to severe dehydration associated with wild-type rotavirus infection.

Electron microscopy remains the gold standard method for diagnosis of rotavirus infection from which the distinctive wheel appearance of the virus can be observed. Due to the cumbersome nature and impracticality of this method, it has essentially been phased out of clinical laboratories. Similarly, culturing rotavirus is no longer routinely performed for clinical diagnosis. Although rotavirus can grow in various cell lines, this process is too time consuming and laborious for meaningful clinical management of disease due to multiple passages being necessary. Instead, diagnosis of rotavirus infection is primarily achieved from stool by antigen EIAs or detection of nucleic acid. Rotavirus antigen EIAs target the VP6 protein most frequently and are highly sensitive and specific. The performance of these tests is not significantly affected by the presence of other enteric bacteria or viruses in stool. However, it is important not to disregard the possibility of co-infection in rotavirus positive specimens or the presence of other bacterial, viral, or parasitic causes of gastroenteritis in cases of similar presentation to rotavirus infection.

FDA-cleared rotavirus PCR assays target rotavirus A, the most common rotavirus group to infect humans, and is included in many of the commercially available, gastrointestinal multiplex panels. Positive results from EIAs or PCR should always be correlated with clinical presentation, as shedding of the virus and free DNA can continue for weeks after primary infection and administration of a rotavirus A vaccine. EIAs are often used in nosocomial outbreak investigations as they are considerably less expensive than PCR, and when an outbreak is documented, the positive predictive value of the assay is very high. Genotyping of rotavirus A is performed at the CDC, where reverse-transcription PCR and sequencing of the VP7 and VP4 genes identify the G and P genotypes, respectively.

Summary Points

- Rotavirus is a non-enveloped, double-stranded RNA virus that is an important cause of childhood gastroenteritis of which the majority of human infections are from rotavirus group A.
- Prolonged illness from rotavirus infection can lead to severe, dehydrating diarrhea in children under 2 years of age.
- Stool antigen detection of the VP6 protein and PCR are the main methods for diagnosis of rotavirus A infection.
- Two rotavirus vaccines are available, a monovalent vaccine containing genotype G1P[8] and a pentavalent reassortant vaccine with genotypes G1 through 4 and P8.

Selected References

Gautam R, Lyde F, Esona MD, Quaye O, Bowen MD. Comparison of Premier™ Rotaclone®, ProSpecT™, and RIDASCREEN® rotavirus enzyme immunoassay kits for detection of rotavirus antigen in stool specimens. J Clin Virol. 2013;58(1):292–4.

Pindyck T, Tate JE, Parashar UD. A decade of experience with rotavirus vaccination in the United States—vaccine uptake, effectiveness, and impact. Expert Rev Vaccines. 2018;17(7):593–606.

Sadiq A, Bostan N, Yinda KC, Naseem S, Sattar S. Rotavirus: genetics, pathogenesis and vaccine advances. Rev Med Virol. 2018;28(6):e2003.

Troeger C, Khalil IA, Rao PC, Cao S, Blacker BF, Ahmed T, Armah G, Bines JE, Brewer TG, Colombara DV, Kang G, Kirkpatrick BD, Kirkwood CD, Mwenda JM, Parashar UD, Petri WA Jr., Riddle MS, Steele AD, Thompson RL, Walson JL, Sanders JW, Mokdad AH, Murray CJL, Hay SI, Reiner RC Jr. Rotavirus vaccination and the global burden of rotavirus diarrhea among children younger than 5 years. JAMA Pediatr. 2018;172(10):958–65.

Case 4

Cameron A. Brown and James J. Dunn

A 14-day-old girl born vaginally at 37 weeks 5 days to a first time mother presents to her pediatrician after her newborn hearing screen found unilateral high frequency sensorineural hearing loss in her right ear as measured by an auditory brainstem response test. The mother's prenatal care was significant for multiple abnormalities on her 20w gestational age (GA) fetal ultrasound, including echogenic bowel patterns. At 32w GA, fetal ultrasound showed polyhydramnios, but otherwise was normal. The mother was healthy throughout pregnancy, but cytomegalovirus (CMV) serology performed in response to the fetal ultrasound was positive for IgM and negative for IgG. Repeat serology to determine IgG seroconversion was not performed. The mother's syphilis test at her first prenatal visit, again at 32w GA, and at delivery were all negative. The mother was prescribed valacyclovir to be taken three times a day until delivery for suspicion of primary CMV infection.

At birth the patient appeared jaundiced and was noted to have darkly colored urine and foul-smelling stool, indicative of neonatal hepatitis and cholestasis. This prompted head and abdominal ultrasounds, with the former finding Grade I intraventricular hemorrhaging, germinal matrix hemorrhaging, and residual germinolytic cysts, and the latter noting mild hepatomegaly. Her hepatitis and cholestasis resolved, and the patient was discharged home for follow-up care with her pediatrician.

At current presentation, complete blood count was largely normal except for a low red blood cell count (3.00 million/μL, Reference Range: 3.10–5.30 million/μL) and elevated mean corpuscular hemoglobin was elevated (40.7 pg, Reference Range: 27.0–36.0 pg). Complete metabolic panel with liver function tests were unremarkable, confirming the resolution of hepatitis. Given the abnormalities observed in utero, hepatitis, hepatomegaly, intraventricular hemorrhaging, and sensorineural hearing loss (SNHL) at birth, and maternal serology results, the chief etiology in the differential diagnosis was CMV. Other etiologies that were considered were focused on organisms that are known to cause congenital infections, such as rubella, herpes simplex virus, toxoplasmosis, and Zika virus. Rubella is known to cause SNHL and hepatic disease in neonates, and Zika virus also frequently causes SNHL though this is generally accompanied by microcephaly and craniofacial disproportion. Other causes of viral hepatitis, including hepatitis A, B, and C, and Epstein-Barr virus, remained on the differential for hepatitis and hepatomegaly. Epstein-Barr virus is known to cross-react with CMV IgM, and thus could have been a false-positive for the maternal serological result.

Because of the strong indication for congenital CMV, urine and saliva from the infant was collected and sent for viral culture. Both specimens were each inoculated into A549 human lung epithelial cell line, human foreskin fibroblasts, and Rhesus monkey kidney cells. After 5 days, multiple enlarged, rounded cells on the human foreskin fibroblasts were observed from both urine and saliva cultures (Fig. 2.8). Confirmation of the virus was achieved with a CMV-specific fluorescent antibody stain on the cell culture material (Fig. 2.9). The patient completed a 6-month course of valganciclovir, and follow-up with audiology showed normal hearing bilaterally using behavioral audiometry.

Fig. 2.8 Human foreskin fibroblast cells inoculated with the patient's saliva. Clusters of enlarged, rounded cells, consistent with the cytopathic effect of cytomegalovirus, can be observed

Discussion

Human cytomegalovirus (CMV) is an enveloped, double-stranded DNA virus in the *Herpesviridae* family of viruses. Many viruses in this family are prevalent in humans, including herpes simplex virus (HSV) 1 and 2, varicella zoster virus (VZV), CMV, and Epstein-Barr virus (EBV). More than half of the population will be infected with CMV by adulthood. The prevalence of infection in childhood is inversely correlated to socioeconomic status. All herpesviruses establish latency in their host after the primary infection, which allows them to persist in the host indefinitely and cause secondary infections upon reactivation later in life. CMV infects epithelial cells, including endothelial cells, and fibroblasts and can spread from cell to cell or induce formation of syncytia. After this primary infection, CMV can establish latency in a variety of tissues and organs, endothelial cells, and leukocytes. Reactivation can occur throughout life but is particularly common in response to immunosuppression.

Infection from CMV commonly occurs through contact with infected bodily fluids from an individual shedding the virus. Body fluids known to contain infectious CMV virions include urine, saliva, semen, vaginal secretions, breast milk, and blood. Vertical transmission is also prevalent, with an estimated 1% of all neonates being congenitally infected. The risk of both symptomatic and asymptomatic congenital CMV infections are higher with a maternal primary infection during pregnancy compared to reactivation from a latent infection. The highest morbidities are observed when the transmission occurs during the first trimester. Perinatal transmission can also occur upon contact with body fluids in the birth canal from a mother shedding virus at the time of delivery or from ingesting breast milk containing the virus. Seronegative recipients of organs, stem cells, or blood from an infected donor can also be infected via transfusion though the latter is rarer today due to leukocyte reduction of blood products.

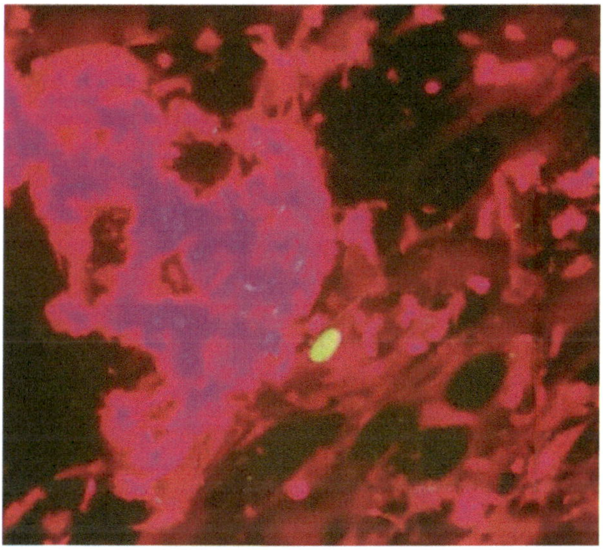

Fig. 2.9 Cells from Fig. 2.8 stained with a fluorescently labeled antibody specific to cytomegalovirus shows apple-green fluorescence, indicating that the cell is infected with cytomegalovirus

In immunocompetent adults and children outside the neonatal period, primary infection is usually asymptomatic. The most common symptomatic presentation is a mononucleosis-like syndrome with prolonged fever, malaise, and myalgia. This presentation can be distinguished from EBV mononucleosis by the lack of heterophile antibodies, and less commonly features pharyngitis and lymphadenopathy. In neonates that acquire the infection perinatally or postnatally, asymptomatic infections still predominate, but a higher proportion will be symptomatic than observed in adults and older children. Neonatal presentation, especially in high-risk individuals such as preterm infants, is often that of lymphadenopathy, hepatitis/hepatosplenomegaly, and pneumonitis. More severe, sepsis-like syndromes and necrotizing enterocolitis have also been associated with perinatal infections.

Congenital CMV is the most common congenital infectious disease and is included in the acronym TORCH (Toxoplasmosis; Others—syphilis, Zika virus, VZV; Rubella; Cytomegalovirus; Herpes simplex virus) for congenital infections that have similar presentations. In utero, CMV findings on ultrasound can include cerebral and hepatic calcifications, microcephaly or abnormal head growth, hepatosplenomegaly, and fetal growth restriction. While 90% of neonates born with congenital CMV are asymptomatic, the most common findings in symptomatic neonates include rash, jaundice, and hepatosplenomegaly. The rash is often a reddish-blue purpura that leads to the description "blueberry muffin baby." Congenital CMV is also the leading cause of sensorineural hearing loss and is present in one third of neonates with symptoms and an estimated 20% of asymptomatic infants will develop this symptom by the time they reach adulthood. Newborns are routinely tested for hearing loss to screen for congenital CMV infections.

CMV can be detected in a variety of tissue and body fluids. In normal or immunocompromised hosts who are at risk of CMV viremia, whole blood and plasma can be used to quantitatively measure the amount of CMV DNA using quantitative PCR (qPCR). For immunocompromised hosts undergoing treatment, a baseline viral load should be obtained prior to starting antiviral therapy, and the amount of CMV can be monitored on a weekly basis during the treatment period to assess treatment efficacy. During this period, the same specimen type, extraction device, performing laboratory and assay should be used to minimize variability. The World Health Organization has an international standard that can be used to calibrate each laboratory's assay to allow for standardization of reporting.

Congenital CMV in the neonate is often diagnosed using saliva or urine. To establish congenital transmission and rule out perinatal transmission, samples should be collected within the first 21 days of life. Historically, viral cell culture or shell-vial assays have been the preferred method, but molecular diagnostics are also now used frequently. Viral culture is best performed using a fibroblast cell line, such as MRC-5. Cytopathic effects (CPE) may be slow to form, taking 10 days or longer, but in specimens with high viral titer, CMV may rapidly proliferate and destroy the cell culture monolayer. Common CPE observed in fibroblast cells include enlarged, rounded, refractile cells. This CPE is difficult to distinguish from other viruses such as adenovirus and VZV, so confirmation of CMV should be determined with a fluorescent antibody specific to CMV. In utero, diagnosis can be made by viral culture or PCR of amniotic fluid or CMV IgM detected in fetal blood.

Due to the high seroprevalence of many adults and children for CMV infections, serology is primarily used for screening at-risk populations or diagnosis of primary infections in children and neonates. However, routine screening of pregnant women for CMV infections is not currently recommended by groups such as American College of Obstetricians and Gynecologists and Society for Maternal-Fetal Medicine. Seroconversion with CMV-specific IgG can be used to establish a primary infection, while IgM can be present during both primary and secondary infections and occasionally persists. IgM detection in a newborn can be useful in determining congenital infection, but false-positives with rheumatoid factor and false-negatives with excess CMV IgG can occur. Because of the limitations in IgM testing, primary infection is often diagnosed using a CMV IgG avidity assay. Following a primary infection, IgG avidity is low, but increases over time. Avidity can be measured by an enzyme immunoassay after denaturing the antigen-antibody complex with urea and comparing the results to a reaction that was not denatured. IgG antibodies produced 6 months or more after infection will have matured and will remain bound after being washed with urea (i.e., high avidity antibodies), while IgG produced early after primary infection will wash off (i.e., low avidity antibodies). A four-fold increase in IgG in paired sera collected 2 or more weeks apart is also suggestive of recent infection but does not distinguish between primary infection and reactivation.

Ganciclovir and its orally available prodrug valganciclovir are common first-line antivirals used for treatment and prophylaxis of CMV infections, and their use is primarily limited to immunocompromised hosts and symptomatic infants with congenital infection. The efficacy of providing antiviral therapy to pregnant women with primary infection in preventing or limiting sequelae in the neonate has not been established. Mutations in the CMV UL97 gene can lead to resistance to ganciclovir and valganciclovir, for which second-line antivirals like foscarnet or cidofovir may be necessary, while mutations in the CMV UL54 gene can lead to resistance to all four of these antivirals. The first indication of treatment failure due to resistance is a persistently stable or increasing viral load or a viral load that recovers after initial response to therapy. In these instances, the relevant genes of CMV can be sequenced directly from blood specimens to determine antiviral susceptibility.

Summary Points

- Cytomegalovirus (CMV) is a DNA virus in the *Herpesviridae* family that can be found in 40–100% of all adults. CMV establishes lifelong latency in a variety of cell types and can reactivate, often secondary to immunosuppression. CMV is spread through contact of body fluids, including oral secretions, congenitally through amniotic fluid, or from a transplant from an infected donor.
- In immunocompetent children and adults, a primary infection with CMV is largely asymptomatic or produces an illness resembling mononucleosis. More severe disease occurs in neonates with congenital transmission and in immunocompromised patients, such as those undergoing transplant or with cellular immunodeficiencies.
- CMV can be diagnosed from a variety of body fluids and tissues. Urine and saliva are commonly used to diagnose neonates for congenital CMV using viral culture or PCR, while quantitative PCR is preferred to monitor CMV DNA levels in blood in immunocompromised patients.
- Ganciclovir and valganciclovir are the preferred antivirals for immunocompromised patients with CMV viremia or neonates with congenital CMV disease. Mutations that confer resistance can be detected using sequencing assays.

Selected References

Fowler KB, Boppana SB. Congenital cytomegalovirus infection. Semin Perinatol. 2018;42:149–54.

Kenneson A, Cannon MJ. Review and meta-analysis of the epidemiology of congenital cytomegalovirus (CMV) infection. Rev Med Virol. 2007;17:253–76.
Manicklal S, Emery VC, Lazzarotto T, Boppana SB, Gupta RK. The "silent" global burdern of congenital cytomegalovirus. Clin Microbiol Rev. 2013;26:86–102.

Chapter 3
Mycology Cases

Audrey N. Schuetz, Marc Roger Couturier, Megan H. Amerson-Brown, Angela Ma, Anisha Misra, and James J. Dunn

A. N. Schuetz (✉)
Department of Laboratory Medicine and Pathology, Mayo Clinic, Rochester, MN, USA
e-mail: Schuetz.Audrey@mayo.edu

M. R. Couturier
Department of Pathology, University of Utah/ARUP Laboratories, Salt Lake City, UT, USA
e-mail: Marc.Couturier@aruplab.com

M. H. Amerson-Brown
Department of Pathology, UAB Heersink School of Medicine, The University of Alabama at Birmingham, Birmingham, AL, USA
e-mail: mamersonbrown@uabmc.edu

A. Ma
Public Health Ontario, Department of Laboratory Medicine and Pathobiology, University of Toronto, Toronto, ON, Canada
e-mail: Angela.Ma@oahpp.ca

A. Misra
Department of Pathology, Cleveland Clinic/Lerner College of Medicine, Cleveland, OH, USA
e-mail: misraa2@ccf.org

J. J. Dunn
Department of Pathology, Texas Children's Hospital, Houston, TX, USA

Department of Pathology & Immunology, Baylor College of Medicine, Houston, TX, USA
e-mail: jjdunn@texaschildrens.org

J. J. Dunn et al., *Challenging Cases in Diagnostic Clinical Microbiology*,
https://doi.org/10.1007/978-3-031-65092-5_3

Overview of Diagnostic Mycology Methods

The fungi comprise a group of spore-forming organisms which feed on organic matter for carbon and energy sources. Mycology is the study of fungi. Fungi are ubiquitous throughout nature as yeasts, molds, mushrooms, and toadstools. They live in soil, in water, and on plants. As opposed to bacteria which are prokaryotic, fungi are eukaryotes, which means that they possess membrane-bound organelles such as nuclei and mitochondria. The fungal cell wall is a characteristic structure of the kingdom. The cell wall is composed predominantly of glycoproteins, chitin, and glucans, which are not present in humans, making it an ideal target for diagnostics and therapeutics.

Fungi produce spores or conidia to propagate and spread as aerosols. They play an essential role in many ecosystems of our planet, as nature's major recycler of organic remains. Many have been adapted by humans to help develop diverse industries, such as food manufacturing (e.g., yeasts in fermentation products such as beer, and molds in cheese-making) and development of pharmaceuticals. In fact, the antibiotic penicillin was originally extracted from the *Penicillium* mold, and researchers continue to develop antimicrobials and other agents from fungi.

Fungal human infections are largely caused by yeasts or molds. Yeasts are unicellular fungi, have a round to oval shape, and typically reproduce by budding. *Candida* are the most common yeasts recovered in the clinical laboratory, but there are many other types of yeasts including *Cryptococcus* and *Malassezia*. Molds are multicellular and composed of filaments called hyphae. Some fungi are dimorphic, meaning that they display either a yeast or mold form, depending on their surroundings. Thermally dimorphic fungi alter their form depending on temperature; this group includes *Blastomyces*, *Histoplasma*, *Sporothrix schenckii*, and *Paracoccidioides*, among others. *Coccidioides* is not dependent upon temperature for its dimorphism. The large group of thermally monomorphic molds is generally divided into Mucorales (also known as the Zygomycetes and includes *Mucor* and *Rhizopus*); dermatophytes (target keratinized tissues); the large group of hyaline hyphomycetes (includes *Aspergillus* and *Fusarium*); and the dematiaceous or dark-pigmented molds. Recent taxonomic studies have classified microsporidia and *Pneumocystis* as fungi; these organisms were previously classified as parasites and are often discussed in older textbooks alongside the parasites.

Fungi can infect any tissue in the human body. Some fungi are commensals, otherwise known as colonizers. Most fungi are opportunists and can lead to disease when a person's immune system is compromised, or when there is a breach in the natural host defenses of the body. For instance, *Candida* yeasts which colonize the mucous membranes can disseminate throughout the body to cause disease in cases of severe neutropenia (i.e., low neutrophil counts). As the number of immunosuppressed individuals continues to rise, fungi that were previously regarded as non-pathogenic are emerging as opportunistic pathogens.

The field of clinical mycology is ever-growing, and the advent of new technologies used for organism identification has transformed the field in the last few years. One such technique is matrix-assisted laser desorption-ionization–time-of-flight mass spectrometry (MALDI-TOF MS), which identifies organisms by their peptide components. Through this technique, peptides within the sample (i.e., fungus) are converted into ions by a laser beam. The ionized peptides are then accelerated through a flight tube; the time required to travel through the tube is determined by the mass and charge of the various ions. A characteristic spectrum or fingerprint of the peptide fragments is generated and compared with mass spectrometry spectra of known microbial organisms contained in a database. The process is rapid, and the method is highly accurate. MALDI-TOF MS is often used for yeast identification, but molds can be identified by this method as well. Highly sensitive molecular assays also provide rapid and accurate identification. However, morphologic examination of cultivated fungi from culture remains the main method of identification of molds in many laboratories. Recovery of the organism for phenotypic antifungal resistance testing is still required for most fungi, as molecular methods of antifungal resistance testing are not widely available.

Specimens

Proper collection technique is critical when collecting specimens for mycologic studies. Since fungi are present throughout the natural environment and as natural microbiota or flora of human skin and mucous membranes, proper specimen decontamination before sampling is essential to avoid detection of commensal or environmental fungi. Swabs should generally be avoided for fungal studies since limited patient material can be gathered on a swab. Rapid transport of specimens to the laboratory is necessary for adequate growth of certain fastidious (i.e., finicky or delicate) fungi. In particular, the hyphae of Mucorales break apart easily when handled. Mucorales hyphae may not survive the process that most laboratories use to grind or crush tissue or biopsy specimens for plating. The provider ordering the culture must note their suspicion for Mucorales when submitting the specimen to the laboratory so that gentle mincing rather than grinding of the tissue will take place, thus increasing the likelihood that Mucorales will grow in culture.

Specimen Processing

Biosafety level (BSL) requirements for handling of specimens for mycologic studies are more stringent than those of most other microbiology laboratory sections that handle bacteria or parasites. Although most yeasts can be manipulated under BSL-2 conditions, cultures of most of the dimorphic fungi should be handled with increased safety precautions in a BSL-3 facility due to higher risk of infectivity from certain fungi.

Specimens submitted for fungal culture are plated to a variety of media types such as general-purpose media (e.g., Sabouraud dextrose agar); selective media (e.g., inhibitory mold agar which contains antibiotics such as chloramphenicol and gentamicin to suppress bacterial growth, or Sabouraud agar with cycloheximide which inhibits growth of contaminating environmental fungi); enriched media (e.g., brain heart infusion agar with blood to enhance growth of dimorphic fungi); media which is both selective and differential (e.g., chromogenic media for *Candida*); and/ or specialized media (e.g., media with olive oil overlay for recovery of *Malassezia*). The choice of agar depends upon the submitted specimen type but nonselective and selective media are generally used for specimen plating.

Examination of Direct Specimens

Microscopic examination of direct patient material for fungal organisms is accomplished using special stains. Direct examination of submitted material is an important part of patient care, as it can provide an initial clue as to whether an organism is present and can aid the treating clinician in narrowing the differential diagnosis. Direct microscopy reports which include specific morphologic terms such as yeasts, hyphae, and pseudohyphae (i.e., an elongated yeast cell with a constriction along its length, much like a chain of sausages) (Fig. 3.1) aid in guiding therapy since treatment varies by the type of fungus.

Potassium hydroxide (KOH) is a rapid stain often used to identify fungal elements in direct specimens. A drop of 20% KOH on a glass slide inoculated with specimen scrapings digests and clears the human tissue, allowing fungal elements to be better visualized under a light microscope. KOH stains are most often used for examination of direct dermatologic material of skin or scalp scrapings to assess for the presence of dermatophytes. Calcofluor white is a fluorescent stain consisting of a nonspecific fluorochrome which binds to cellulose and chitin in the fungal cell wall; fungal elements will appear bright bluish/white when examined under the fluoresence microscope using an excitation filter of 380–420 nm and emission filter of 450 nm. Immunofluorescence stains for *Pneumocystis* are also often used. India ink was used

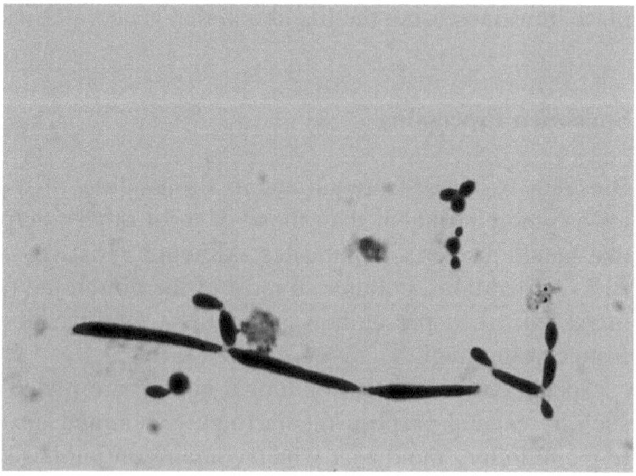

Fig. 3.1 Gram stain of abdominal fluid demonstrates oval yeasts and elongated, pinched-off pseudohyphae of *Candida albicans*, much like a chain of sausages (400× magnification)

in the past to detect *Cryptococcus* in cerebrospinal fluid (CSF) and other specimens, but antigen detection has replaced it due to low sensitivity of India ink.

Culture

Macroscopic and microscopic appearances of fungi growing on media have long been the sole mainstay of fungal diagnosis for molds but are more recently becoming paired with more sophisticated methods such as MALDI-TOF MS or molecular assays. Yeast identification is most often achieved by MALDI-TOF MS or commercial identification cards that assess biochemical characteristics of the yeasts; however, morphologic identification can also be a supporting identification method for yeasts. Sequencing-based identification is used less frequently for yeasts but may be necessary in some instances.

Yeasts are the most common fungi recovered in culture in the clinical laboratory. They often grow on culture plates as glabrous (i.e., smooth) colonies without apparent aerial hyphae or projections (Fig. 3.2). Yeasts may be subcultured to certain agars such as cornmeal agar for examination of their microscopic morphology to aid in identification.

Molds, on the other hand, appear fuzzy, powdery, or fluffy on media, much like mold growing on old bread which has been sitting at room temperature too long (Fig. 3.3). Colony colors often provide a helpful clue for the organism. For instance, *Aspergillus fumigatus* colonies are typically blue-green, while *A. niger* colonies are black. The presence of unique fungal elements microscopically is key, however, for identification of molds. Unique structural characteristics of molds such as the size, shape, and color of conidia or spores, as well as the hyphae themselves (i.e., presence of septa, type of branching, diameter of hyphae) aid in morphologic characterization of molds.

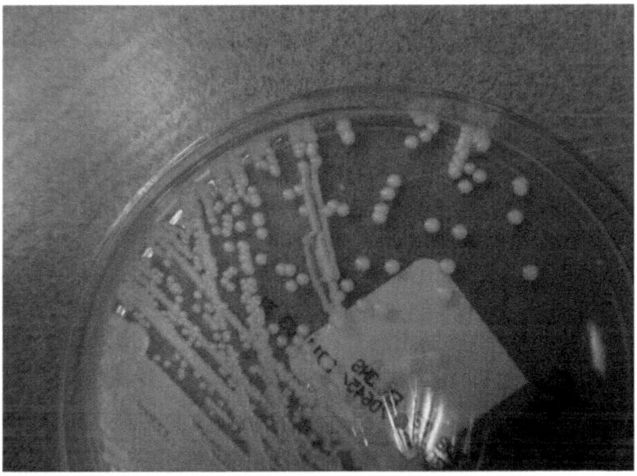

Fig. 3.2 Glabrous *Candida albicans* colonies on Sabouraud dextrose agar

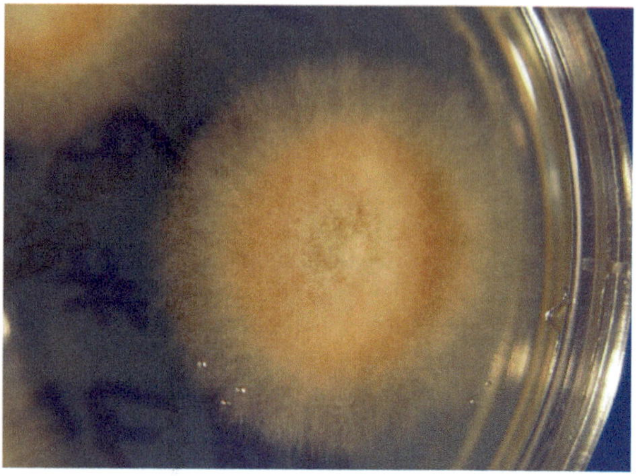

Fig. 3.3 Orange-colored, fuzzy, fluffy *Fusarium* mold colonies on Sabouraud agar

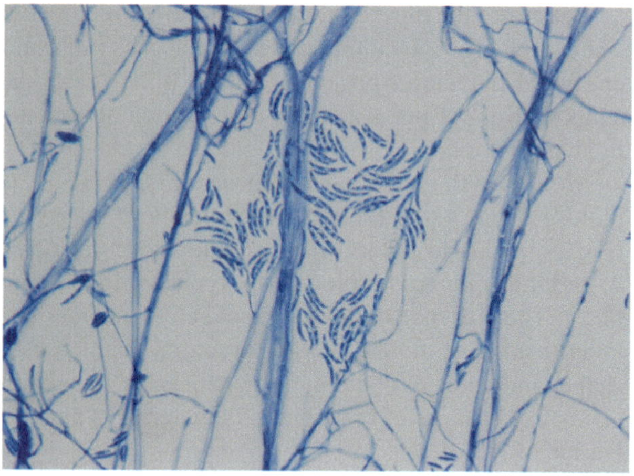

Fig. 3.4 Lactophenol cotton blue tape preparation of *Fusarium* colony, showing hyphae and characteristic sickle-shaped conidia (400× magnification)

Fungi growing on culture plates can be visualized microscopically by preparing a tape preparation of the colony growth. With this method, cellophane (clear) tape is pressed firmly against the fungal colony. The tape is then adhered to a glass slide and then stained with lactophenol cotton blue or aniline dye to highlight fungal structures (Fig. 3.4). Alternatively, a tease mount can be performed by placing a small portion of the colony onto a glass slide and gently teasing apart the colony with dissecting needles, followed by staining with either lactophenol cotton blue or aniline dye. A slide culture is the preferred method for preserving structural integrity of the mold for microscopic examination but is difficult to perform entirely in a biosafety cabinet and thus should not be set up if thermally dimorphic molds are suspected.

Histopathology

Many fungi display unique morphologic characteristics when examined histologically in formalin-fixed paraffin-embedded (FFPE) tissues or on cytologic examination. The type of surrounding tissue reaction (e.g., formation of granulomas or neutrophilic inflammation) provides further clues as to the fungus present. The appearance of certain fungi in tissue may differ drastically from the microscopic appearance on culture plates. For instance, thermally dimorphic fungi appear as yeast forms in human tissue (at 37°C, which is normal human body temperature) but as mold forms in nature or on culture plates incubated in the laboratory at 30°C. Thus, the dimorphic fungus *Histoplasma* is a small, budding oval yeast on histopathologic examination (Fig. 3.5), while microbiologists examining colonies of *Histoplasma* grown from culture plates visualize hyphae with spiny microconidia (Fig. 3.6).

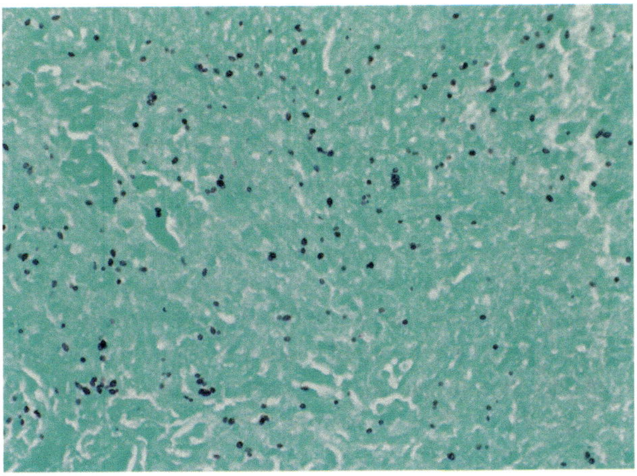

Fig. 3.5 Histopathology Gomori methenamine silver (GMS) stain showing many small, budding oval yeasts of *Histoplasma* (400× magnification)

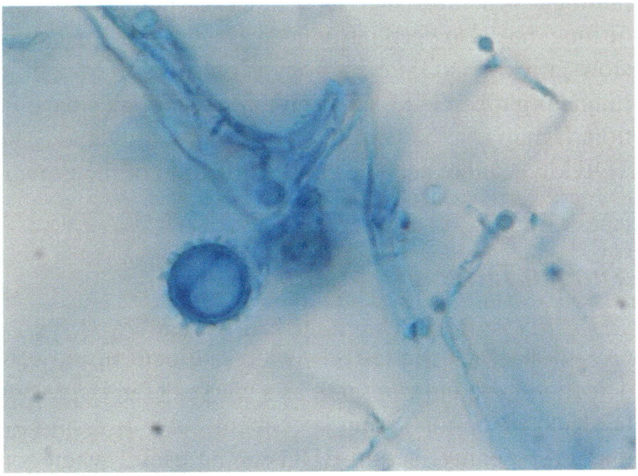

Fig. 3.6 Lactophenol cotton blue tape preparation of *Histoplasma*, showing characteristic large, spiny macroconidium and a background of hyphae (1000× magnification)

Although many fungi can be seen on hematoxylin and eosin (H&E) staining, occasionally fungi may be difficult to see without special stains. Grocott's or Gomori methenamine silver (GMS) stain highlights fungal elements in black on histopathology (Fig. 3.5). The periodic acid-Schiff (PAS) stain with or without diastase is an alternative to GMS and highlights fungi red-magenta. Mucicarmine and Alcian blue stain the capsule of *Cryptococcus*. The melanin stain Fontana-Masson colors dematiaceous fungi black, as well *Cryptococcus* which has melanin in its cell wall.

Molecular and Proteomic Methods

MALDI-TOF MS, PCR, and DNA sequencing of fungi have revolutionized laboratory identification of fungi in the last decade. Many laboratories rely primarily on MALDI-TOF MS for identification of yeasts. Increasingly, laboratories also rely on this method for mold identification. Currently, yeasts and dimorphic pathogens are well-represented in the commercial MALDI-TOF MS libraries, whereas many molds are not; however, these databases are constantly being updated, and diagnostic accuracy has greatly improved since the development of the technology.

In general, target-specific PCR may be performed direct from patient specimens or on isolates for identification. Direct patient specimens include either fresh (i.e., unfixed) patient material submitted to the clinical microbiology laboratory, or FFPE. Molecular methods complement morphologic studies and are highly accurate. Many laboratories run target-specific PCR assays for endemic mycoses such as *Histoplasma*, *Blastomyces*, and *Coccidioides*, as well as *Pneumocystis jirovecii*. Sanger sequencing targeting the internal transcribed spacer (ITS) and/or the D1/D2 regions has also been used to identify known fungi and aid in the discovery of new ones. Local population characteristics and laboratory size and volume drive the need for specific molecular testing offered in-house or sent to reference laboratories. Sanger sequencing of fungi is useful when a yeast is not identified by biochemical commercial methods or if a mold is not sporulating (i.e., does not display usual morphologic structures such as conidia or spores). Next-generation sequencing (NGS) of fungi has also been introduced as a tool for fungal diagnosis since some fungi are slow-growing. NGS technology is most helpful for fungal diagnosis in cases of low fungal organism load, or when other techniques have failed to lead to an identification. It may also be helpful when morphology and Sanger sequencing (or MALDI-TOF MS) cannot achieve an identification of the organism to the species level.

Antigen Detection Methods

(1,3)-Beta-D-glucan (BDG) is a constituent of the cell wall of most fungi, and its detection in serum can serve as a marker of invasive fungal infection. BDG can also be detected in non-serum specimens such as respiratory sources. As a pan-fungal marker, a positive serum BDG can be useful when considering empiric antifungal therapy until a specific organism can be identified. BDG is secreted in high amounts during infection with *Pneumocystis*. Fungi that lack BDG include *Cryptococcus*, *Blastomyces*, and the Mucorales.

Galactomannan is a major component of the cells wall of *Aspergillus*. Serum and bronchoalveolar lavage galactomannan levels are helpful in ruling in or out aspergillosis and other fungal diseases in certain patient populations.

Serologic Testing

Several serologic tests exist for the dimorphic fungi. Oftentimes, a diagnosis of disease due to dimorphic fungi relies upon a combination of assays such as serologic testing combined with culture, antigen, and/or molecular testing. While some serologic tests are qualitative, it is the quantitative assays which provide antibody titers or levels and may be helpful in both diagnosis and follow-up of disease course in a patient. Some serologic assays assess IgM and some IgG, while others do not differentiate among antibody class. Fungal diagnosis by serology often depends on a combination of various serologic tests (e.g., enzyme-linked immunoassays [EIAs], immunodiffusion antibody tests, complement fixation). For instance, serologic testing for coccidioidomycosis includes measurement of IgM and IgG EIA first, followed by immunodiffusion antibody testing if the EIA is positive. *Coccidioides* EIAs are sensitive, but IgM EIA is less specific; therefore, a positive EIA should be followed by the highly specific immunodiffusion test for confirmation and complement fixation for trending titers over time. Serologic testing is also helpful in establishing a diagnosis of blastomycosis and histoplasmosis.

Antimicrobial Susceptibility Testing

Antimicrobial susceptibility testing (AST) can be performed for yeast and molds; however, AST for molds is often performed only at specialized reference laboratories due to the complexity of testing. AST of yeasts is generally performed using broth microdilution by a commercial method. Breakpoints and interpretive guidelines for AST results are generally limited to *Candida* and *Cryptococcus* for certain antifungal agents, and only for a few select molds. Additional studies and data are necessary to improve this important gap in fungal infection therapeutics.

Summary

Clinical mycology is a growing field, with unique safety risks associated with certain pathogens. Analytical detection processes are challenged by the ubiquitous nature of fungi. A combination of methods such as direct examination, culture, histopathologic appearance, molecular/proteomic methods, antigen detection, and serologic testing is often required to reach a diagnosis. Antifungal susceptibility testing is likewise expanding to include testing of organisms beyond *Candida* species. The development and use of molecular methods will push the field of clinical mycology forward.

Case 1

Marc Roger Couturier

A 43-year-old farmer from eastern Idaho presented to the emergency department with a 3-day history of corneal pain, photophobia, redness of the sclera, and swelling of the conjunctiva in his left eye. The patient noted a left eye corneal abrasion that he sustained approximately 7 days prior to admission. He worked as a potato farmer and was recently in the field turning over diseased potato crops for the purpose of incineration. While in the field, his colleague that was flying a crop duster flew very closely over his head while applying antifungals to the remaining crops, causing dust, dirt, and debris to get into his left eye. He immediately attempted to flush his eye with irrigation water but felt a retained "pebble" or other object which he was not able to remove until later that evening. He recalled that his eye felt scratched when he blinked his eyelid.

Bacterial and fungal pathogens were both of concern to the emergency department physicians. Emergency ophthalmology was consulted, and a corneal abrasion was confirmed. Corneal scrapings were collected and submitted for aerobic culture and fungal culture. Due to the soil exposure, the physicians recommended empiric therapy using eyedrops, including tobramycin and 5% natamycin. Bacterial cultures were negative for growth at 72 h, but the fungal media began to grow white/cream colored aerial mycelia (Fig. 3.7). The growth was sampled for MALDI-TOF mass spectrometry but failed to yield an identification. Two days later, there was more abundant growth, and a tape preparation was made and stained with lactophenol cotton blue and material was again extracted for MALDI-TOF. The tape preparation revealed small oval-shaped microconidia with one or two cells and large macroconidia with 3–5 septations and a "canoe-like" fusiform morphology (Fig. 3.8). The phenotypic identification was made as *Fusarium* sp. The repeat MALDI-TOF testing yielded an identification of *Fusarium solani* complex. The patient was told to discontinue the tobramycin, continue the 5% natamycin, and was also prescribed 1% topical voriconazole for 2 weeks.

When the patient was told the organism he was being treated for, he immediately recognized the name. The patient has been removing the diseased crops due to identified *Fusarium* leaf wilt, a common infection of potato crops in Idaho. The crop-duster likely stirred up the mold spores, and they became inoculated into the eye with the dirt and soil. The abrasions in the cornea from the debris likely allowed the *Fusarium* to superficially invade the cornea. The patient recovered without any complications and only minimal corneal scarring.

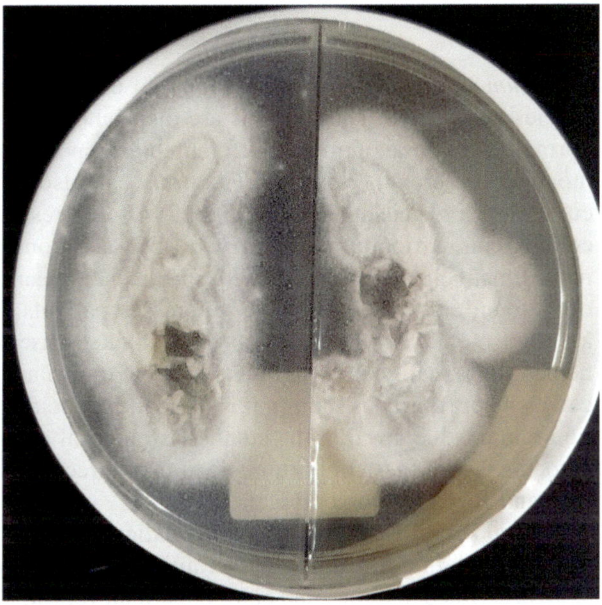

Fig. 3.7 *Fusarium solani* complex growth on Sabouraud dextrose agar (left) and potato dextrose agar (right) bi-plate. Colonies on both agars show areas of removed colony growth for MALDI-TOF mass spectrometry and microscopic examination

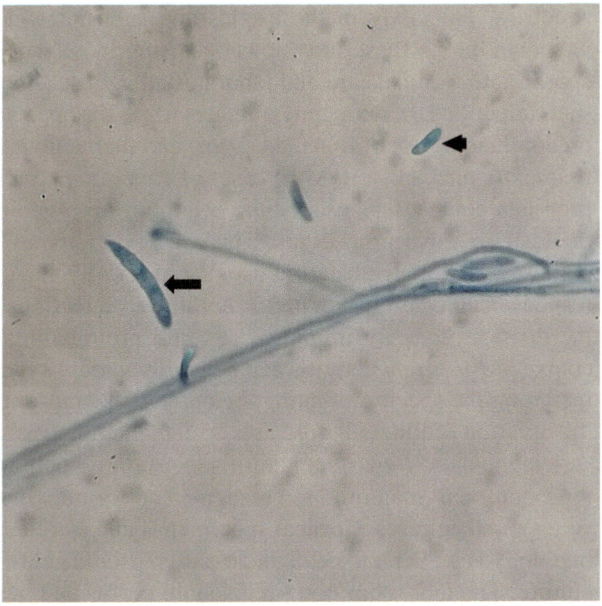

Fig. 3.8 Microscopic analysis of tape preparation from solid agar growth. Hyphae, microconidia (arrowhead), and macroconidia (arrow) can all be seen in this field. Image captured at 400× magnification

Discussion

Fusarium is a genus of filamentous mold that contains over 300 species, and 20 species complexes. These molds are distributed worldwide and are known to be normal mycoflora of many different plant species (especially mass-produced, monoculture crops). However, *Fusarium* can also be pathogenic to some monocultural crops, causing *Fusarium* leaf wilt and resulting in significant loss of to commercial agricultural. *Fusarium* spp. are considered opportunistic fungal pathogens. Human infections with *Fusarium* appear to be increasing, as they are often found in severely immune suppressed patients. Immune suppressed patients are also more likely to develop disseminated disease, especially patients with hematologic malignancy and bone marrow transplant recipients. Healthy patients can also get fusariosis though it is typically associated with trauma and the infection remains localized at the cutaneous/mucosal site of the trauma. Most species associated with these opportunistic human infections belong to the *F. solani* complex and *F. oxysporum* complex.

Fungal keratitis can be difficult to predict from bacterial keratitis based solely on examination. In general, fungal keratitis is statistically more likely to have raised slough, serrated margins, and color other than yellow. *Fusarium* keratitis is a rapidly destructive infection and a common cause of monocular blindness in developing countries, especially in the tropics and subtropics. The true incidence is largely unknown due to these infections occurring largely in countries where prompt and appropriate medical care and laboratory diagnoses may not be readily available for such infections. In temperate climates (e.g., Europe, North America), yeasts are the predominant cause of fungal keratitis, and filamentous molds like *Fusarium* are less frequently encountered. Most cases of *Fusarium* keratitis are associated with inappropriate contact lens use, ocular trauma from organic materials, ocular surgeries, or other invasive procedures. These associations are all made higher risk when paired with prolonged steroid use. Paradoxically, having an overtly robust immune response can cause prolonged, complicated infections in *Fusarium* keratitis due to the damage done by immune cells and proinflammatory cytokines. The clinical course of *Fusarium* keratitis can be highly variable and depends on multiple factors including the species infecting the cornea, the antifungal susceptibility profile, and the duration of illness before appropriate therapy was started.

Diagnosis of *Fusarium* keratitis is primarily achieved through fungal stains, *in vivo* confocal scanning microscopy (performed in specialized ophthalmology suites), and cultures. Corneal specimens can be collected for both routine stains in histology (i.e., hematoxylin and eosin, periodic acid-Schiff) and fungal stains and culture. The most common fungal stain performed in the microbiology laboratory is the calcofluor white stain that is a fluorescent stain that binds cellulose and chitin in biological materials. Identification is not possible through this stain; however, it can be used as a rapid and sensitive screen for fungal elements. The amount of corneal material collected may limit the number of cultures or stains that can be prepared. The clinical probability of fungal versus bacterial keratitis is paramount in situations in which the specimen is limited. Corneal swabs are generally not acceptable for fungal culture though they may be used for bacterial cultures as needed. Fungal cultures may not yield growth in many cases of true fungal keratitis, so empiric therapy may be continued even in the absence of culture growth if clinical suspicion is high. *Fusarium* spp. generally grow rapidly on Sabouraud dextrose agar incubated at 25 °C and will initially appear as white, flat, and cottony growth (Fig. 3.7). The colonies may develop other colors on the surface after a few days of growth, including red, pink, purple, salmon, and tan/brown. The reverse is typically tan, red, purple, or brown, with early growth often showing no color. The microscopic features of *Fusarium* includes septate hyphae, 2–5 × 8–16 μm oblong/pyriform/oval microconidia with 1–3 septations, and 4–6 × < 65 μm macroconidia with 3–5 septations (Fig. 3.8). Identification can be achieved to the species-complex level by ITS and D1/D2 sequence analysis, whereas species-level identification typically requires multilocus gene sequencing. MALDI-TOF mass spectrometry can reliably identify *Fusarium* to the species-complex level, but not reliably to the species level.

Fusarium represent one of the most antifungal resistant genera of molds. *Fusarium solani* is typically considered the most resistant of all the species of *Fusarium*. Amphotericin B may have relatively low MICs, whereas flucytosine, fluconazole, ketoconazole, miconazole, itraconazole, and posaconazole all typically have high MICs. Voriconazole typically has lower MICs than the other azole drugs. *Fusarium* is intrinsically resistant to all echinocandins though some evidence of synergistic treatment with amphotericin has been reported. *Fusarium* keratitis is typically treated with topical natamycin at 5% concentration plus or minus 1% topical voriconazole. Treatment is aggressive, typically with these two topical antifungals being administered every hour for the first 48 h and every 2 h for 1–2 weeks thereafter. There is no established standard length of antifungal duration in these patients, and in the case of corneal grafts, treatment may be extended for months.

Summary Points

- *Fusarium* is a genus of filamentous fungi that are normal plant flora and plant pathogens but are also capable of infecting humans; both immune competent and immune compromised.
- *Fusarium* spp. typically have high MICs to most antifungal agents, With the exception of amphotericin B which can be variable.
- *Fusarium* will typically grow rapidly in culture and can be readily identified to genus level by characteristic microscopic features of the macroconidia.
- MALDI-TOF mass spectrometry and ITS sequencing can be used to identify *Fusarium* to the species-complex level, while multilocus sequencing is required for a species-level identification.
- *Fusarium* keratitis is a common cause of fungal keratitis in tropical and subtropical countries, but is also found in temperate climates with less frequency.

Selected References

Nucci M, Anaissie E. *Fusarium* infections in immunocompromised patients. Clin Microbiol Rev. 2007;20(4):695–704.

Sleiman S, Halliday CL, Chapman B, Brown M, Nitschke J, Lau AF, Chen SC. Performance of matrix-assisted laser desorption ionization-time of flight mass spectrometry for identification of *Aspergillus, Scedosporium*, and *Fusarium* spp. in the Australian clinical setting. J Clin Microbiol. 2016;54(8):2182–6.

Szaliński M, Zgryźniak A, Rubisz I, Gajdzis M, Kaczmarek R, Przeździecka-Dołyk J. *Fusarium* Keratitis—review of current treatment possibilities. J Clin Med. 2021;10(23):5468.

Case 2

Megan H. Amerson-Brown and James J. Dunn

A 16-month-old female with lacrimo-auriculo-dento-digital syndrome (LADDS) and failure of lung development was transferred from an outside hospital to a specialty children's hospital for evaluation of management and a lung transplant. LADDS is a genetic disorder characterized by defects in the tear-producing system, ear problems, dental abnormalities, and deformities of the fingers. The patient presented with pulmonary hypotension, pulmonary hypoplasia, and renal agenesis with a singular right kidney. At the time of transfer, the patient was ventilator dependent and her primary nutrition source was total parenteral nutrition (TPN). She had previously completed 12 weeks of voriconazole for antifungal treatment due to increasing serum galactomannan levels. Upon discontinuation of antifungal therapy, the patient had new recurring fevers and a rising serum galactomannan index equal to 4.13 (reference range <0.5) with unclear etiology. A chest X-ray was performed. The results were clinically consistent with tracheitis and ventilator-associated pneumonia (Fig. 3.9). Three days after discontinuing voriconazole, the patient decompensated developing bradycardia and apnea. General labs revealed a high WBC count (15,000/μL) with neutrophilia (58%) and lymphopenia (22%). The patient also had hypernatremia (161 mmol/L), hypokalemia (3 mmol/L), hyperglycemia (166 mg/dL), and a high lactic acid (1.9 mmol/L). The patient had a femoral central venous line in place for 93 days and a peripherally inserted central catheter (PICC) line in place for 119 days. Upon decompensation, aerobic, anaerobic, and fungal blood cultures were obtained and sent to the microbiology laboratory.

Three days after culture collection, the aerobic blood culture bottle flagged positive. The Gram stain showed wide based unipolar budding yeast from the aerobic blood culture bottle (Fig. 3.10). The blood culture bottle was subcultured to blood agar, chocolate agar, MacConkey agar, and Sabouraud dextrose agar (SDA). The subculture had no growth after 72 h. The fungal blood culture was also negative. Suspecting that additional

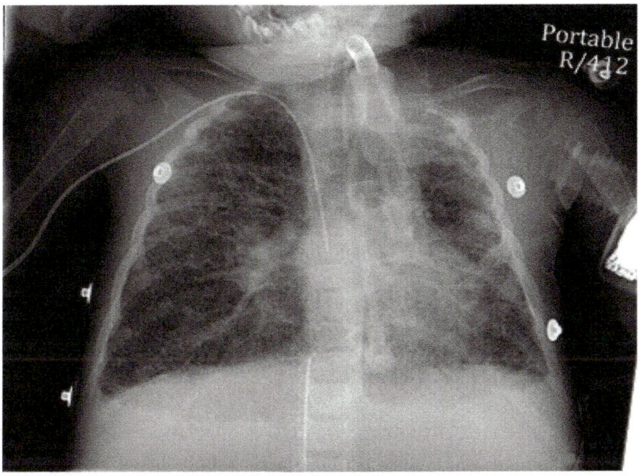

Fig. 3.9 Chest X-ray

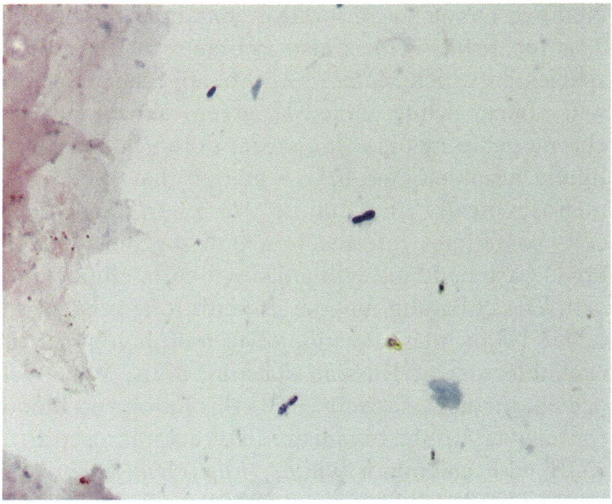

Fig. 3.10 Blood culture Gram stain (magnification, 1000×)

nutritional requirements might be needed for the yeast to grow, the SDA plate was reset from the blood culture bottle with an overlay of sterile olive oil. White, smooth colonies grew only where the oil was present. *Malassezia furfur* was identified by MALDI-TOF. The patient's central line was replaced and the patient was treated with amphotericin B for 10 days. All subsequent blood cultures were negative.

Discussion

The genus *Malassezia* includes 14 species, most of which are lipid dependent, with the exception of *M. pachydermatis*. *Malassezia* is a common commensal of the human skin and hair follicles. The most common clinical manifestations of

Malassezia infections include various skin manifestations such as superficial mycosis, pityriasis (i.e., tinea) versicolor, seborrheic dermatitis, dandruff, atopic eczema, and folliculitis. The most common species isolated from clinical specimens include *M. furfur, M. pachydermatis, M. sympodialis, M. globosa, M. obtusa, M. restricta,* and *M. slooffiae.* Fungemia due to lipid-dependent *Malassezia* species is associated with high mortality rates and most frequently occurs in immunosuppressed infants and young children with central line catheters receiving lipid replacement therapy or parenteral lipids. *Malassezia* is less frequently isolated from immunosuppressed adults receiving TPN. Common symptoms of *Malassezia* fungemia infection include fever, bradycardia, apnea, thrombocytopenia, and catheter blockage. *M. pachydermatis* displays similar clinical manifestations as lipid-dependent *Malassezia* spp.

The most common isolation technique used for *Malassezia* spp., except *M. pachydermatis,* is to overlay SDA with olive oil or use a media containing long chain fatty acids. Specialized media such as Leeming and Notham agar, modified Dixon's agar, or agar containing Tween 80 can be used. Chromogenic agar for *Malassezia* is also available for the primary isolation of *Malassezia* species, in which *Malassezia* spp. appear pink or pink-purple while other yeast will appear white, blue, or green. When cultured on SDA or other non-chromogenic media, *Malassezia* colonies can be cream to yellow, smooth or lightly wrinkled, and have a margin that is either entire or lobate. When seen microscopically from culture, *Malassezia* appears as globose, ellipsoidal yeast cells sometimes referred to a bowling pin. Reproduction occurs by unipolar broad base budding with a distinct collarette between the mother and daughter cell. The collarette may be difficult to appreciate unless the slide is examined under 1000× magnification. The morphology of the budding yeast is said to resemble that of Russian stacking dolls. When *Malassezia* fungemia is suspected, the use of special fungal isolation and blood culture systems are useful for recovery of the organism. In direct microscopy from skin scrapings, such as KOH with calcofluor white, *Malassezia furfur* may appear as variable yeast forms with round budding yeast-like cells and short hyphal forms often referred to as a "spaghetti and meatball" appearance (Fig. 3.11).

There are currently no FDA-approved rapid nucleic acid-based assays for identification of yeast in positive blood cultures that will identify *Malassezia* spp. The cell wall of *Malassezia* contains galactomannan and can result in positive galactomannan antigen results. *Malassezia* spp. also contain very small amounts of (1,3) β-D-glucan, but is unlikely to give a positive result when using that biomarker to detect fungal infections.

Treatment of *Malassezia* fungemia includes removing the infected catheter, discontinuing parenteral lipids and treatment with amphotericin B or azole antifungals, including itraconazole or voriconazole. In most cases, the infected catheter must be removed since the organism is not able to be cleared from lipid deposits in the line. For treatment of topical skin infections, topical or oral azoles including ketoconazole or fluconazole are effective.

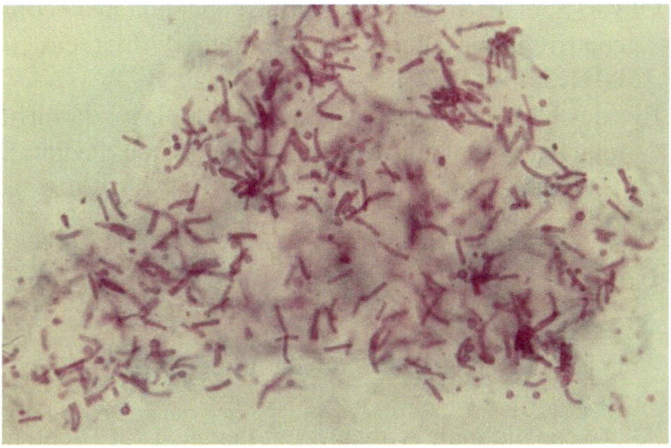

Fig. 3.11 Gram stain of *M. furfur* from skin sample showing characteristics "spaghetti and meatball" appearance (magnification, 1000×). (Available at the CDC Public Health Image Library: https://phil.cdc.gov/Details.aspx?pid=2916)

Summary Points

- *Malassezia* is a unique genus of yeast that can cause topical and systemic infections including fungemia.
- Patients receiving parenteral lipids are at higher risk for infections caused by *Malassezia* than other patients.
- *Malassezia* should be considered in the evaluation of febrile children who are receiving total parenteral nutrition with lipids and who present with immunosuppression, central venous catheters, and persistent fever that is unresponsive to antibiotics.
- *Malassezia* should be considered when yeast is seen on a direct smear or a blood culture, but fails to grow under normal growth conditions. Isolation of most *Malassezia* spp. requires media with added lipids or an oil/lipid overlay after inoculation.
- Treatment for *Malassezia* fungemia includes catheter removal and administration of antifungals, most commonly amphotericin B.

Selected References

Arendrup MC, et al. ESCMID and ECMM joint clinical guidelines for the diagnosis and management of rare invasive yeast infections. Clin Microbiol Infect. 2014;20 Suppl 3:76–98.

Gaitanis G, et al. The malassezia genus in skin and systemic diseases. Clin Microbiol Rev. 2012;25:106–41.

Iatta R, et al. In vitro activity of two amphotericin B formulations against Malassezia furfur strains recovered from patients with bloodstream infections. Med Mycol. 2015;53(3):269–74. https://doi.org/10.1093/mmy/myu089.

Oliveri S, et al. Malassezia furfur fungaemia in a neonatal patient detected by lysis-centrifugation blood culture method: first case reported in Italy. Mycoses. 2011;54(5):e638–40.
Rosales CM, Jackson MA, Zwick D. Malassezia furfur meningitis associated with total parenteral nutrition subdural effusion. Pediatr Dev Pathol. 2004;7(1):86–90.

Case 3

Angela Ma and Marc Roger Couturier

A 24-year-old male with a recent diagnosis of B-cell acute lymphoblastic leukemia (B-ALL) admitted for induction chemotherapy was evaluated by a dermatologist regarding a painful bone marrow biopsy site. The bone marrow biopsy was performed on his left iliac area 2 weeks prior, and the lesion was initially thought to be a small hematoma with cellulitis. He was prescribed doxycycline for 7 days but failed to show improvement. The patient now complained of worsening pain and concerns for localized skin necrosis. Upon examination, it was noted that the lesion had enlarged and darkened to a 3 cm by 3 cm black eschar surrounded by a deep purple, faintly reticulated border (Fig. 3.12). Superficial cutaneous thrombosis was highest on the dermatology team's differential, but it was necessary to rule out infectious etiologies. Two 4 mm punch biopsies were obtained for histopathology and bacterial, fungal, and mycobacterial culture. No growth was observed on all cultures, but histopathology identified broad, pauciseptate fungal hyphae branching at obtuse angles that was suggestive for mucormycosis.

Upon obtaining these results, the patient was shortly scheduled for debridement of the left iliac lesion with plastic surgery. He also received an infectious

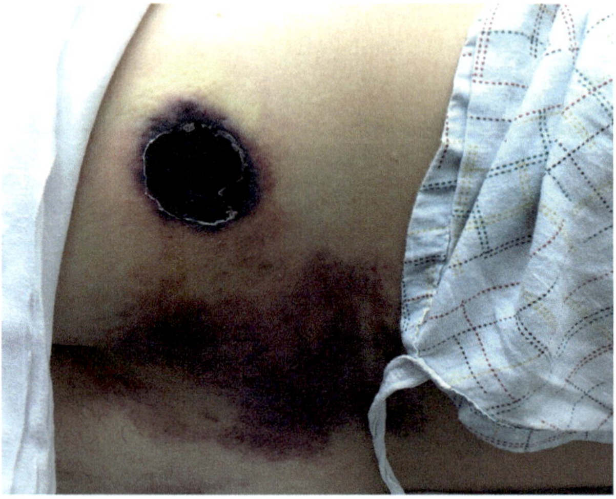

Fig. 3.12 Black eschar with reticulated purple edges correlating with cutaneous mucormycosis in the left iliac region of an immunocompromised patient with B-cell acute lymphoblastic leukemia

disease consult and was started on intravenous amphotericin B. Four days after the patient's initial punch biopsies were collected, the lesion had grown to measure 10 cm by 12 cm in size and was tender to touch and indurated. From his debridement procedure, ten specimens were collected; frozen sections were performed on three fresh specimens and six were sent for histopathology in formalin. From the remaining debridement specimen collected, microbiological cultures were inoculated. Due to the high probability of fungal infection, a calcofluor white with potassium hydroxide (KOH) stain on tissue, serum (1,3)-beta-D-glucan test, *Aspergillus* species PCR test on tissue, and Mucorales PCR test on tissue were also ordered. Two sets of fungal blood cultures were collected to determine if the patient had fungemia.

Histopathology demonstrated extensive necrosis with infiltrating fungal organisms present in four of the nine specimens received (Fig. 3.13). Angioinvasion was noted in two of the subcutaneous tissue specimens, but clear margins were identified from deeper specimens. Negative results were received for the (1,3)-beta-D-glucan test and *Aspergillus* species PCR but the Mucorales PCR was positive. The calcofluor white with KOH stain failed to demonstrate any fungal elements. *Staphylococcus epidermidis* was isolated from the bacterial culture and a gray-brown mold grew within 6 days of inoculating the fungal culture. *Mucor* spp. was identified phenotypically by cellophane tape, lactophenol cotton blue stain preparation. Antifungal susceptibility testing showed elevated minimum inhibitory concentrations (MICs) for isavuconazole and voriconazole but low MICs for amphotericin B, itraconazole, and posaconazole. The mycobacterial culture and both sets of fungal blood cultures yielded no growth. Chest and lung imaging also showed no evidence of sinus or pulmonary involvement. The patient continued treatment with amphotericin B with therapeutic drug monitoring for nephrotoxicity and was also prescribed intravenous daptomycin to cover the *S. epidermidis*. After 3 weeks of therapy, the patient was switched from amphotericin B to posaconazole and he underwent a follow-up debridement with successful wound closure.

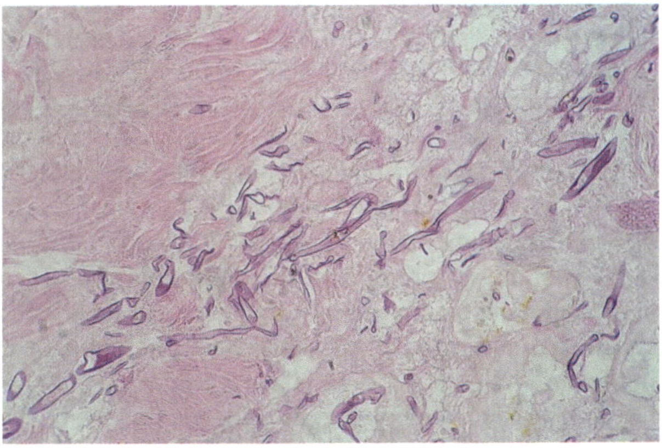

Fig. 3.13 Hematoxylin and eosin-stained subcutaneous tissue section demonstrating broad hyphae suggestive of mucormycosis

Discussion

Mucor spp. belong to the order Mucorales that comprise a large proportion of the mucormycetes (previously known as zygomycetes). Other members of the Mucorales that are commonly implicated in clinical infection include the *Rhizopus* and *Rhizomucor* genera. Twelve out of 76 known species of *Mucor* have been reported to cause human infection with *M. circinelloides* being most frequently isolated. Mucormycosis is characterized by suppurative necrosis of tissue and is typically associated with cutaneous, pulmonary, and rhinocerebral involvement. The mucormycetes also have the capacity for angioinvasion that can lead to increased risk of thrombosis and infarction. The molds are common environmental contaminants found in soil, plants, and decaying vegetation, so correlation with clinical presentation and immune status is important when they are isolated. Patients at risk for mucormycosis include those who are immunosuppressed or neutropenic, individuals with poorly controlled diabetes mellitus, and chronic steroid use among other factors. Mucorales are transmitted primarily by inhalation of the spores into the lungs and paranasal sinuses or direct inoculation into the skin from sites of trauma.

Specimen processing of primary specimens for fungal culture largely resembles bacterial culture, except for tissue specimens. Unlike bacterial cultures that typically require ground tissue to be plated onto culture media, tissue specimens should be minced using a sterile scalpel and placed onto the media surface for fungal cultures. Grinding of tissue may damage fungal structures and impair recovery of viable organisms. Enriched and nutrient-deficient media are regularly employed for primary isolation of molds and induction of sporulation for phenotypic identification, respectively. The media can be supplemented with antibacterial agents including chloramphenicol, gentamicin, and cycloheximide but some members of the Mucorales, including *Mucor* spp., are inhibited by cycloheximide. The Mucorales are commonly referred to as "lid-lifters" due to the ability of some members to fill the entirety of the culture plate or tube with aerial mycelia within a few days of incubation. They grow on solid media as white to gray, wooly or "cotton candy-like" molds that darken with age (Fig. 3.14). The reverse side of *Mucor* spp. and other Mucorales on solid media is typically white.

There are distinct morphologic features specific to the different members of the Mucorales when examined microscopically. Overall, the Mucorales possess broad, pauciseptate hyphae that are often described as "ribbon-like." The sporangiophores of *Mucor* spp. and *Rhizomucor* spp. can be branched, whereas in *Rhizopus* spp. they are typically singular or in unbranching groups. An apophysis may or may not be present depending on the genus. In *Mucor* spp., the apophysis is absent. At the end of the sporangiophore is the asexual reproductive structure of the Mucorales known as the sporangium, from which spores are released. *Rhizopus* spp., *Rhizomucor* spp., and *Mucor* spp., all possess round sporangia of similar size. A key distinguishing feature to the Mucorales is the presence or absence of rhizoids, root-like structures along the hyphal base. Rhizoids are present in *Rhizopus* spp. and *Rhizomucor* spp. but are absent in *Mucor* spp. (Fig. 3.15).

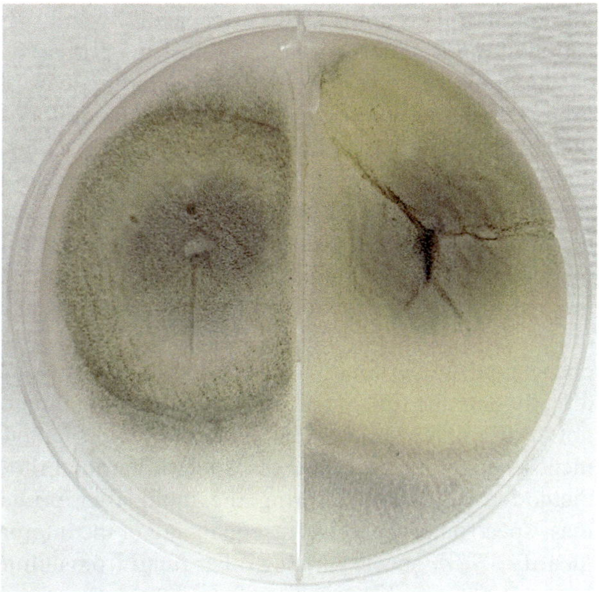

Fig. 3.14 *Mucor* spp. on potato dextrose agar (left) and Sabouraud dextrose agar (right) showing floccose mycelia

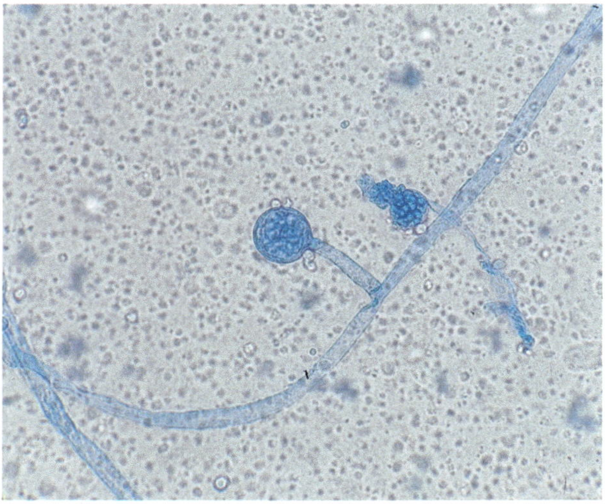

Fig. 3.15 Cellophane tape preparation of *Mucor* spp. stained with lactophenol cotton blue showing broad, nonseptate hyphae without rhizoids and a sporangiophore lacking an apophysis at 40× magnification

Diagnosis of *Mucor* spp. and other molds can be achieved phenotypically through identification of macroscopic and microscopic features. A common method for direct microscopic examination of specimens for fungal elements is the calcofluor white with KOH stain. Calcofluor white is a nonspecific fluorochrome with an affinity for cellulose and chitin in fungal cell walls and fluoresces blue-white at 400 nm. When combined with KOH, the alkaline solution serves as a clearing agent to reduce fluorescence by non-fungal material. However, one limitation of the stain is its lack of sensitivity and specificity. Phenotypic identification of fungi is most successful following primary isolation from culture. Molds can be prepared by various methods including tease and cellophane tape preparations, which can be rapidly set up in the laboratory, and slide culture. All three methods use lactophenol cotton blue, which is comprised of lactic acid, phenol, and the cotton blue stain. Lactic acid functions similarly to KOH, clearing the specimen of background material. Phenol serves as an inactivation agent to render the mold noninfectious, and cotton blue stains chitin in the cell wall to provide increased contrast of fungal elements. One limitation of the tease preparation is the inability to preserve the original orientation and structure of fungal elements due to physical manipulation of the mycelium. This is reduced with the cellophane tape preparation, which involves pressing the sticky side of clear tape directly onto the fungal colony and transferring it onto a glass slide with lactophenol cotton blue. Slide culture is the optimal method for visualizing fungal elements, but significantly increases the turnaround time for identification since the method requires subculturing the mold after primary isolation. MALDI-TOF mass spectrometry and DNA sequencing of the internal spacer regions 1 and 2 can identify *Mucor* species among other fungi from culture when phenotypic identification is not possible.

Currently, no serologic or antigen detection tests for *Mucor* spp. or other Mucorales are available for clinical diagnosis. Although the serum (1,3)-beta-D-glucan test is a fungal antigen detection assay that can aid in diagnosis of disseminated fungal infections, it cannot detect members of the Mucorales because they do not produce (1,3)-beta-D-glucan. Some reference laboratories offer Mucorales PCR assays that detect the most common members of the order from culture or direct patient specimens. These assays generally target a shared sequence of the 18S rRNA gene of the Mucorales. While the PCR assays do not provide species-level identification, a positive result can significantly aid in tailoring therapy. Azole and echinocandin class antifungals have decreased to no activity against the Mucorales; amphotericin B is the main antifungal agent for the treatment of mucormycosis. Furthermore, there are currently no guidelines for the interpretation of antifungal susceptibility testing results for the Mucorales.

In cases of invasive fungal disease, histopathology is the gold standard. Histopathology of tissue biopsies can suggest involvement of the Mucorales due to their distinctive broad hyphae that can be differentiated from the narrow, septate hyphae of hyaline molds. In hematoxylin and eosin-stained tissue sections, Mucorales hyphae often have a "crinkled cellophane" appearance. Although special stains such as the Grocott's methenamine silver (GMS) stain are useful for identifying fungal elements in tissue, the hyphae of the Mucorales may stain poorly with them. Ancillary tests that can be performed from fresh and formalin-fixed, paraffin-embedded tissue include broad-range PCR and DNA sequencing assays that are offered by specialized laboratories.

Summary Points

- *Mucor* spp. and other members of the Mucorales are ubiquitous environmental molds that are naturally found in soil and decomposing organic matter.
- The Mucorales are capable of angioinvasion and their broad, nonseptate hyphae can be visualized in histopathology of tissue and from fungal culture.
- *Mucor* spp. differ from *Rhizopus* spp. and *Rhizomucor* spp. morphologically due to the lack of an apophysis and rhizoids.
- MALDI-TOF mass spectrometry and molecular PCR and DNA sequencing assays can be used for detection and identification of *Mucor* species.
- There is no antigen detection or serology tests available to diagnose mucormycosis.
- Treatment is typically amphotericin B due to resistance or decreased susceptibility to the echinocandin and azole antifungals.

Selected References

Chandra S, Woodgyer A. Primary cutaneous zygomycosis due to mucor circinelloides. Australas J Dermatol. 2002;43(1):39–42.

Petrikkos G, Skiada A, Lortholary O, Roilides E, Walsh TJ, Kontoyiannis DP. Epidemiology and clinical manifestations of mucormycosis. Clin Infect Dis. 2012;54 Suppl 1:S23–34.

Rammaert B, Angebault C, Scemla A, et al. Mucor irregularis-associated cutaneous mucormycosis: case report and review. Med Mycol Case Rep. 2014;6:62–5.

Walsh TJ, Hayden RT, Larone DH. Mucormycetes. In: Larone's medically important fungi: a guide to identification. 6th ed. Washington, DC: ASM Press; 2018:183–201.

Case 4

Megan H. Amerson-Brown and James J. Dunn

A 12-year-old male with a history of ulcerative colitis and moderate persistent asthma, currently being treated with adalimumab, a TNF-alpha blocker, presented to the emergency department at an outside hospital. The patient was complaining of a 10-day history of fever (103 °F), chest pain, and hypoxia. The patient lived in rural East Texas on a farm where multiple livestock animals were raised including chickens, goats, and cattle. Prior to this admission, he was seen by a primary care provider and was given amoxicillin for potential *S. pyogenes* (Group A Streptococcus) pharyngitis, as confirmed by a positive rapid streptococcal antigen test from a throat

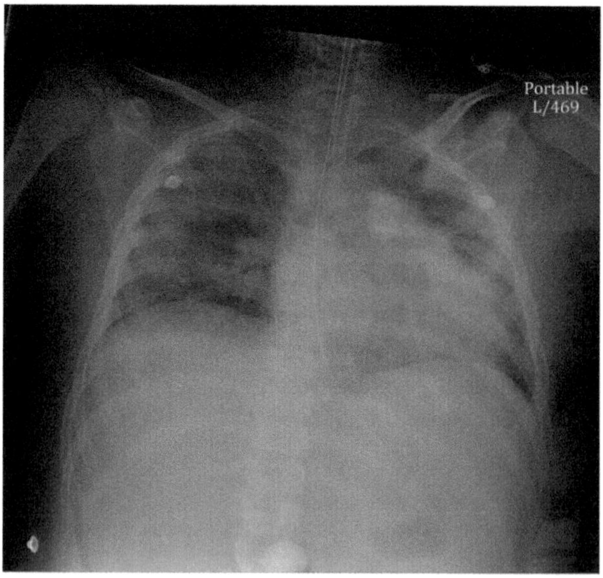

Fig. 3.16 Chest X-ray performed at the outside facility revealing diffuse opacities suggestive of bilateral pneumonia

swab. The patient did not show improvement after treatment with amoxicillin. Chest X-ray (Fig. 3.16) performed at the outside hospital revealed bilateral peribronchial pneumonia. Lymphadenopathy of the mesenteric and right lower quadrant lymph nodes were also noted on an abdominal CT. His blood work revealed an elevated WBC = 24,100/μL (44% neutrophils and 16% bands). He had low levels of platelets and hemoglobin with elevated lactate dehydrogenase, C-reactive protein, erythrocyte sedimentation rate, ferritin, D-dimer, liver enzymes, and bilirubin indicating a severe inflammatory response. CSF was obtained by a lumbar puncture and Gram's stain, culture, and a meningitis/encephalitis PCR panel for common bacterial and viral central nervous system pathogens was negative. Imaging of the brain was unimpressive, suggesting there was no CNS involvement. The patient developed a rash during his admission at the outside hospital that started on the upper back and extended to the rest of his body (Fig. 3.17a, b).

After an extensive workup of coagulation studies and SARS-CoV-2 serology to rule out disseminated intravascular coagulation and multisystem inflammatory syndrome in children, respectively, the patient was diagnosed with hemophagocytic lymphohistiocytosis (HLH) thought to be caused by an infectious process. At this time, the patient was also found to have elevated liver enzymes and hyperbilirubinemia with concerns for liver impairment. He was transferred from the outside hospital to the children's specialty hospital, intubated and sedated. Treatment with broad-spectrum antibiotics (cefepime, metronidazole, and vancomycin) and antifungals (micafungin) was initiated. At this time, the differential diagnosis included endemic dimorphic fungi such as *Coccidioides*, *Histoplasma*, and *Blastomyces*, and

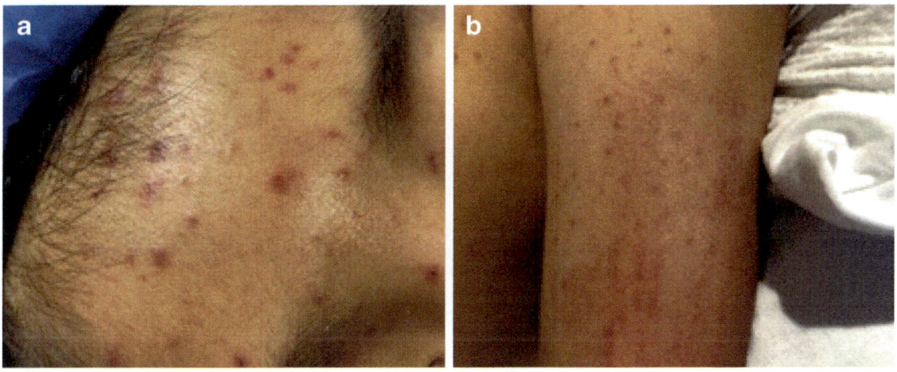

Fig. 3.17 Images of the full-body petechial rash that appeared after 10 days of symptoms. (**a**) Forehead (**b**) left arm

CMV, HSV, Adenovirus, EBV, *Pneumocystis jirovecii* pneumonia, and mycobacteria infections (including *Mycobacterium tuberculosis*), murine typhus, Ehrlichiosis, and Leptospirosis. A second lumbar puncture was performed to collect CSF for endemic dimorphic fungi serology and culture as well as routine bacterial culture. Blood cultures and a bronchoalveolar lavage fluid were also collected for microbiology testing. Additional samples were collected for serology testing, galactomannan, and viral PCR testing. Imaging of the head, chest, and abdomen at this time revealed bilateral pneumonia and abnormalities relating to hepatomegaly with diffuse steatosis, borderline splenomegaly, and thickening of the gallbladder likely secondary to the hepatic process. An ocular exam was also performed as the care team was initially suspicious for possible ocular involvement, but the lesions resolved with no ocular impairment.

Serology performed on the CSF sample for endemic dimorphic mycosis detected *Blastomyces* CSF antibody (reference range is no antibody detected). However, the *Blastomyces* urine antigen and blood antigen were both negative. Fungal compliment fixation tests were all negative. All other PCR and serology testing for viral etiologies were negative. All bacterial and mycobacterial cultures were negative for pathogens. Cytopathology from the right upper lobe bronchoalveolar lavage revealed many intracellular clusters of small yeast on a Grocott's methenamine silver stain (GMS). A punch biopsy of the rash skin showed mild predominantly superficial dermal perivascular lymphocytic infiltrates with scattered yeast forms measuring 3–5 μm (Fig. 3.18). The *Histoplasma* antigen was also positive from the urine, serum, and BAL. The serum galactomannan index was reported as 2.97 (reference range <0.05). Fungal cultures from the respiratory tract grew a white cotton mold on Sabouraud dextrose agar (SDA) that was identified as *Histoplasma capsulatum* (Fig. 3.19) and confirmed by DNA probe. The patient was successfully treated with amphotericin B.

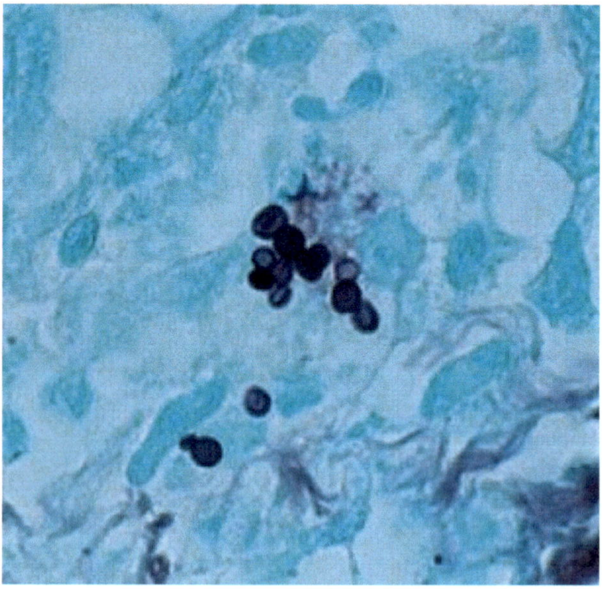

Fig. 3.18 GMS stain (1000× magnification) of a skin punch biopsy showing small round yeast measuring 3–5 µm in diameter

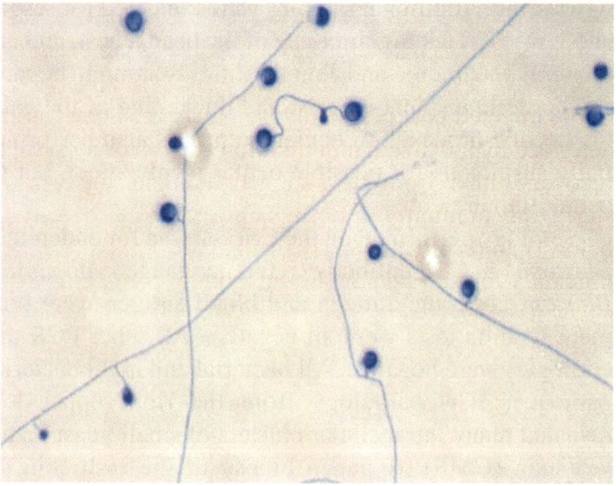

Fig. 3.19 Lactophenol blue stain (400× magnification) of *H. capsulatum* prepared by tape prep from a respiratory culture

Discussion

There are two varieties of *H. capsulatum*, var. *duboisii* and var. *capsulatum*. *H. capsulatum* var. *duboisii* is primarily found in central and western Africa, while *H. capsulatum* var. *capsulatum* can be found worldwide. Historically, in the United States, *H. capsulatum* has been recognized as a cause of endemic mycoses in the Mississippi-Ohio River Valley. Due to climate change, the boundaries for endemic histoplasmosis have expanded and currently encompass the eastern US coast and extend west encompassing most of the central United States including Texas, Oklahoma, Kansas, Nebraska, and the Dakotas. Infections typically occur through inhalation of the fungi from disrupted soil or old structures. Occupational exposure such as construction or activities that disrupt the soil, especially soil that has bat or bird excrement, is considered high risk for acquisition of histoplasmosis in endemic areas.

Histoplasmosis causes a wide spectrum of clinical manifestation. Common symptoms include fever, headache, nonproductive cough, pleuritic chest pain, and fatigue. Most infections (95%) are asymptomatic or self-limiting transient pulmonary infections. Severe infections can cause chronic progressive lung disease, chronic cutaneous infections, or disseminated disease. Disseminated disease is most commonly seen immunocompromised patients and can be fatal if left untreated. Lymphatic, splenic, hepatic, and adrenal gland involvement is common in disseminated disease. Although uncommon, excessive immune activation as seen in HLH has been triggered by histoplasmosis further complicating and increasing the severity and mortality of the disseminated infection. Mucosal lesions or a cutaneous rash can also occur with disseminated infections. CNS involvement occurs in 10–20% of infections in immunocompromised persons. Treatment is usually amphotericin B followed by prolonged suppressive therapy of itraconazole if the patient remains immune suppressed.

Histoplasma capsulatum is a dimorphic fungus that takes on characteristics and structures of hyaline filamentous fungi in the environment and yeast at body temperature. When isolated in the laboratory at 25 °C, the colonies in the mycelial phase are typically white to tan, slow-growing cottony colonies, with the reverse being yellow to brown in color. Isolation of the mold can take 2–6 weeks to grow on routine fungal culture media such as Sabouraud dextrose agar. Microscopically, they will have septate hyphae with microconidia (1–5 μm in diameter) (Fig. 3.19) and tuberculate macroconidia (8–14 μm in diameter) (Fig. 3.6) that arise from lateral and terminal branches. At 37 °C, the fungi take on characteristics of yeast. The yeast form will be seen from direct specimens and when grown at higher temperatures on brain heat infusion agar. Microscopically, the yeast measures 3–5 μm and is typically seen in macrophages on direct smears or in tissue. The cells can be described as single budding daughter cells with a narrow base. Occasionally, these can appear similar to the amastigotes of *Leishmania* spp. Commercially available molecular probes can be used to confirm isolates suspected of being *Histoplasma capsulatum*. These assays utilize the ability of complementary nucleic acid strands to specifically align and associate to form a stable double stranded. When the probe is hybridized to the organism's ribosomal RNA, the chemiluminescent label is detected. A positive result is a reading of the signal greater than or equal to a cutoff value. A negative result is less than the cutoff value.

The sensitivity of laboratory testing for the detection of *H. capsulatum* depends on the clinical manifestation, burden of disease, and in the case of serology, time since exposure. The sensitivity of culture in patients with Histoplasmosis is 42–74%, with sources such as BAL and biopsies being more sensitive than routine blood culture. For disseminated infections, lysis-centrifugation tubes for the recovery of fungi are more sensitive than conventional blood culture methods. *Histoplasma* antigen testing can be performed using an enzyme immunoassay to detect the *Histoplasma* polysaccharide antigen in urine, serum, or CSF. This test has relatively high sensitivity (95%) in disseminated disease. However, the specificity is low since cross-reactivity can be seen in other types of infections such as those with *Blastomyces* spp., *Coccidioides* spp. and less commonly *Sporothrix* spp. or *Paracoccidioides* spp. *Histoplasma* complement fixation tests uses serum to detect antibodies to *H. capsulatum*. In most assays, titers ≥1:32 are suggestive of *H. capsulatum* infection. Antibodies for *H. capsulatum* can cross-react with *Blastomyces* and *Coccidiodes*. Antibody levels for histoplasmosis can also be low in the CSF leading to false negatives. Serum or BAL galactomannan as well as 1,3-β-D-glucan can be positive in histoplasmosis infections but are not sensitive or specific for *H. capsulatum* and therefore must be confirmed with *Histoplasma*-specific assays.

Summary Points

- Most infections of *H. capsulatum* are subclinical with only symptomatic or severe disease occurring in immunocompromised persons or the very old or young.
- *H. capsulatum* var. *capsulatum* is endemic in the Eastern part of the United States extending from the Atlantic coast through the Great Plains.
- *H. capsulatum* is a dimorphic fungi that grows in a mycelia form at room temperature and in the environment and as a yeast for in the body at temperatures ≥37 °C.
- Culture is the gold standard for clinical identification of *H. capsulatum*. Clinical isolates can take up to 6 weeks to grow in the mycelial form and can be identified by their tuberculate macroconidia. Confirmatory testing using DNA probes can confirm the identification of *H. capsulatum*.
- Many biomarkers and serological tests are available for *H. capsulatum*; however, sensitivity and specificity of the tests is dependent upon disease time, manifestation, and severity. Cross-reactivity between other fungi is not uncommon.

Selected References

Azar MM, Hage CA. Laboratory diagnostics for histoplasmosis. J Clin Microbiol. 2017;55(6):1612–20. https://doi.org/10.1128/JCM.02430-16.

Jabr R, El Atrouni W, Male HJ, Hammoud KA. Histoplasmosis-associated hemophagocytic lymphohistiocytosis: a review of the literature. Can J Infect Dis Med Microbiol. 2019;2019:7107326. https://doi.org/10.1155/2019/7107326.

Linder KA, Kauffman CA. Histoplasmosis: epidemiology, diagnosis, and clinical manifestations. Curr Fungal Infect Rep. 2019;13:120–8.
Wheat LJ, Freifeld AG, Kleiman MB, Baddley JW, McKinsey DS, Loyd JE, Kauffman CA. Clinical practice guidelines for the management of patients with histoplasmosis: 2007 update by the Infectious Diseases Society of America. Clin Infect Dis. 200;45(7):807–25.

Case 5

Marc Roger Couturier

A 31-year-old male with a 1-year history of intermittent diarrhea, tenesmus, and incontinence presented to his primary care physician with acute progressively worsening diarrhea. He also recently experienced nausea, vomiting, chills, and significant weight loss. The patient lived in Salt Lake City, UT, for 15 years and was originally from Kingston, Jamaica. His past medical history was significant for human immunodeficiency virus (HIV), which had not been well controlled. The patient had also been previously diagnosed with syphilis, which was appropriately treated, and hepatitis B which had not been treated. Over the course of the previous year, he had been recommended by the primary care physician to submit stool for enteric pathogen workup on numerous occasions, but the patient had declined to submit specimens due to the continual resolution of symptoms. The patient noted that he received an abdominal CT from an outside hospital and was told it was "unremarkable," but those records were not available.

The patient was advised by his primary care physician to seek attention in the emergency department at a local hospital. On admission, the patient denied any recent travel or exotic/unusual diet. He recalled new sexual partners, both male and female. He endorsed having unprotected sex including receptive anal sex and anilingus as well as occasional intravenous drug use and daily marijuana use. A repeat abdominal CT scan showed mild diffuse colonic wall thickening with fluid within loops of small and large bowel compatible with enterocolitis. Workup included a multiplex stool pathogen PCR panel which detected enteropathogenic *E. coli* of questionable significance given the prolonged nature of his clinical course. Stool cultures were also negative for *Shigella, Salmonella, Campylobacter,* shiga-like toxin-producing *E. coli, Aeromonas, Plesiomonas,* and *Vibrio*. The patient was started on bictegravir/emtricitabine/tenofovir alafenamide for his HIV, ceftriaxone for unclear indications, and intravenous fluids with some improvement. Additional laboratory testing was obtained to interrogate his previous diagnoses of HIV and hepatitis B. The hepatitis B viral load was 6.1 log IU/mL, CD4 T-cell count was 111 cells/μL, and HIV viral load was 6.5×10^5 copies/mL. Stool was collected and submitted for parasite examination including a routine ova and parasite examination consisting of a concentrated wet mount and trichrome stain, a modified acid-fast stain, and Ryan's modified trichrome stain. Ova and parasite examination and

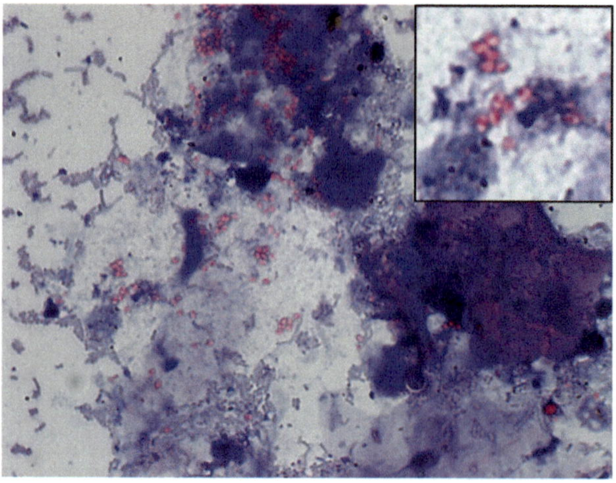

Fig. 3.20 *Encephalitozoon* spp. stained with modified trichrome Ryan's blue method. Numerous spores measuring approximately 2–3 µm are seen in faint pink, with some spores having darkly pink-stained equatorial bands (enlarged in inset) (Image captured at 1000× magnification)

modified acid-fast stain for coccidia and *Cryptosporidium* were negative. The modified trichrome stain was positive for microsporidia spores with numerous spores per high power field of view (Fig. 3.20). To determine the genus of the infecting microsporidia, and in an effort to guide therapy, a microsporidia PCR was ordered. The PCR can detect and distinguish *Encephalitozoon* spp. (*E. intestinalis*, *E. hellem*, or *E. cuniculi*) from *Enterocytozoon bieneusi*. The PCR was positive for *Encephalitozoon* spp. The patient was prescribed albendazole, and in combination with antiretroviral therapy, his symptoms resolved quickly and was discharged.

Discussion

Members of the phylum Microsporidia are obligate intracellular, unicellular, spore-forming fungi. The phylum was previously considered to be comprised of primitive protozoa, however with additional genome sequencing and phylogenetic analysis, the phylum was reclassified to the kingdom Fungi. Microsporidia have a membrane-bound nucleus and a polar tubule/polar filament, which is a unique organelle coiled inside the spore. The polar tubule is everted from the germinated spore and acts as a syringe to inject the infective sporoplasm into host cells. The sporoplasm proliferates within the "parasitized" host cell in a parasitophorous vacuole where it undergoes fission. The parasitophorous vacuole eventually ruptures, releasing infective spores.

The phylum Microsporidia consists of approximately 1400 species capable of infecting both vertebrates and invertebrates. Approximately 200 genera have been described with at least 15 species identified as opportunistic human pathogens though this could be an underestimate. Marked increases in clinical cases have been reported since the emergence of HIV and chemotherapeutic-induced immunosuppression. *Enterocytozoon bieneusi* is considered to be the primary human pathogenic microsporidia, causing significant diarrhea and even death in susceptible hosts. The genus *Encephalitozoon* contains three species that are also associated with opportunistic infection (*E. cuniculi, E. hellem, E. intestinalis*). Other less frequently reported genera and species, and those not typically associated with gastrointestinal infection include *Anncaliia, Microsporidium, Nosema ocularum, Pleistophora ronneafiei, Tubulinosema acridophagus, Trachipleistophora*, and *Vittaforma corneae*. These species are typically associated with ocular, CNS, skin/soft tissue, and large organ infections.

Microsporidia are ubiquitous in the environment due to their environmentally resilient spores, which are highly resistant to desiccation, ultraviolet light, and heat. The spores can be spread through dust and can be dispersed across many biotic and abiotic surfaces. Due to the broad dispersion of microsporidia, incidental ingestion of spores is considered the most significant route of acquisition. Infection with microsporidia has been documented in immunocompetent and asymptomatic individuals; however, microsporidiosis is considered an opportunistic infection of patients with compromised immune systems. The most frequently reported host population is HIV-infected patients with poorly managed disease. Other immunocompromised hosts can also be infected with microsporidia, including solid organ transplant or bone marrow transplant recipients receiving monoclonal antibody therapy. The clinical manifestations can be highly variable depending on the location of infection and the species causing the infection. These symptoms can include diarrhea, keratoconjunctivitis, cellulitis, hepatitis, peritonitis, urethritis/prostatitis, and rarely fatal disseminated disease. Diarrhea is the most common manifestation, which can be caused by *E. bieneusi* and *Encephalitozoon* spp. Both *E. bieneusi* and *E. intestinalis* are capable of infecting enterocytes in the small intestine, and as a result these organisms account for the majority of microsporidia-associated diarrheal disease.

The gold standard for detection and identification of microsporidia is transmission electron microscopy; however, this is not practical for clinical use and is rarely employed in a diagnostic setting. The most common approach to detect microsporidia in stool is visualization of spores by light microscopy using modified trichrome stained, concentrated, fixed stool. Two variations of the modified trichrome stain include the Chromotrope 2R—Ryan's blue method and modified trichrome—Weber green method. Spore walls will stain a weak pink color (Fig. 3.20) with darker staining equatorial bands or "belts" seen at the midpole of the spore (Fig. 3.20, inset). The spores range in size from 1 to 4 μm, which overlaps with both bacteria and yeast. Careful examination of the morphology, stain pattern, and intensity is required to accurately interpret these stains. Calcofluor white stain can also be used with stool specimens, but it can be difficult to interpret since it is a nonspecific stain that binds any structures containing chitin or cellulose. Calcofluor white stain is easier to interpret on ocular fluids, urine, and CSF specimens due to reduced background material and the sterile nature of the specimens.

Molecular methods have also been developed for detection of microsporidia from clinical specimens, although these organisms are not included in most commercially available panels. Unlike stool protozoa, the spores of microsporidia are very difficult to lyse and extract nucleic acid. Physical bead-beating followed by chemical extraction of nucleic acid is required for optimal sensitivity. Assays can be designed to include genus- or species-specific primers for conserved gene sequences in these organisms. These assays are sensitive for detection and typically designed to detect the most common human gastrointestinal pathogenic species (i.e., *E. bieneusi* and *Encephalitozoon spp.*). Distinguishing these genera can have important treatment implications, as *Encephalitozoon* spp. generally respond to therapy with albendazole, whereas *E. bieneusi* does not. For patients with intestinal infection with *E. bieneusi*, reduction in immunosuppression (either through proper HIV treatment or reducing immunosuppressive therapies) can allow for spontaneous resolution of disease.

Summary Points

- Microsporidia is a broadly grouped taxonomic phylum of fungi that can be opportunistic pathogens in humans.
- *Enterocytozoon bieneusi* and *Encephalitozoon* spp. account for the majority of microsporidia-associated gastrointestinal infections.
- Patients with significantly compromised immune systems are at highest risk for microsporidial infections, especially HIV patients with poorly managed disease and patients with inherited or chemotherapeutically-induced immunodeficiency.
- Light microscopy is the most widely used method for detection of microsporidia. Stool is typically stained with a modification of the trichrome stain such as Ryan's blue method or Weber-green method.
- PCR may be helpful to distinguish *E. bieneusi* from *Encephalitozoon* spp., which can have important implications in guiding therapy.

Selected References

Centers for Disease Control and Prevention. DPDx—laboratory identification of parasites of public health concern—microsporidiosis. https://www.cdc.gov/dpdx/microsporidiosis/index.html. Accessed 3 Mar 2022.

Han B, Weiss LM. Microsporidia: obligate intracellular pathogens within the fungal kingdom. Microbiol Spectr. 2017;5(2):FUNK-0018-2016.

Wadi L, Reinke AW. Evolution of microsporidia: an extremely successful group of eukaryotic intracellular parasites. PLoS Pathog. 2020;16:e1008276.

Weber R, Deplazes R, Mathis A. Chapter 131. Microsporidia. In: Carroll KC, Pfaller MA, editors. Manual of clinical microbiology. 12th ed. Washington, DC: ASM Press; 2019. p. 2305–16.

Case 6

Anisha Misra and Audrey N. Schuetz

A 52-year-old man with a medical history notable for tobacco use (17 pack year) and diabetes mellitus type 2 presented to the emergency department with fever, cough productive of yellow-green sputum, headaches, myalgias, and sore throat for 1 week. He also reported worsening of his symptoms at night with chills, fatigue, night sweats, and a weight loss of about 5 pounds in the past 2 weeks. A chest radiograph demonstrated lingular airspace opacity concerning for bacterial pneumonia, and he was started on cefdinir and azithromycin. After a week of treatment, his symptoms continued with exertional dyspnea and intermittent tachypnea of over 30 breaths/min which prompted further evaluation. A computed tomography (CT) scan revealed a consolidative opacity involving the left upper lobe lingular segment with additional consolidative opacities in the left lower lobe and enlarged mediastinal lymph nodes.

Due to the patient's persistent symptoms, a bronchoscopy was performed to obtain bronchoalveolar lavage (BAL) fluid to evaluate for pulmonary mycosis. Bacterial and mycobacterial stains and cultures from the BAL were performed. The BAL fluid was cloudy with 13.0 white blood cells per microliter, was negative on acid-fast modified auramine O fluorescent stain (AFB) and Gram stains, but was positive on fungal smears with calcofluor white stain for broad-based budding yeasts (Fig. 3.21). At this time, serum specimens were drawn to evaluate for *Blastomyces* antibodies by enzyme immunoassay and immunodiffusion. A urine antigen test was also sent out for *Blastomyces*; both serum and urine tests were positive. Twelve days later, fungal cultures were positive for *Blastomyces* species complex, identified by macroscopic and microscopic morphology and confirmed by matrix-assisted laser desorption ionization–time-of-flight mass spectrometry

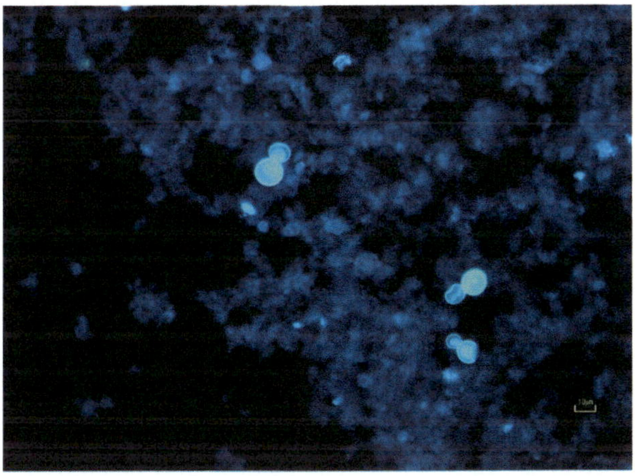

Fig. 3.21 Calcofluor white stain of yeast forms of *Blastomyces*. Large (8–15 μm) yeast cells with broad-based budding and refractile double-contoured cell walls are present

(MALDI-TOF MS). The patient was started on itraconazole for treatment of pulmonary blastomycosis, with plans to continue itraconazole for 6–12 months with a repeat CT scan and *Blastomyces* urinary antigen test at 6 months.

A more detailed history revealed that the patient had been working on building a cabin in northern Minnesota with his wife and dog. They had been excavating the foundation of an old cabin and noticed significant mold and mildew in the air. He had also been sanding fiberglass with a loosely fitting mask and initially believed that to be the reason for his symptoms. One week prior to the patient's presentation, his dog also became ill with severe cough, fever, and dyspnea. The dog was taken to the veterinary clinic where he was found to have severe hypoxemia requiring supplemental oxygen by high-flow nasal cannula. Unfortunately, the dog decompensated further and passed away shortly thereafter. Blastomycosis can progress quite rapidly in humans too and can be fatal; however, in this case the patient's symptoms improved, with a decreasing trend in *Blastomyces* urinary antigen levels over time.

Discussion

Blastomycosis is an infection caused by a fungal pathogen of the genus *Blastomyces*. The etiologic agents of blastomycosis are thermally dimorphic fungi—*Blastomyces dermatitidis* and *Blastomyces gilchristii*. These are environmental pathogens that are typically found in moist soil and decomposing organic matter. *Blastomyces* is a mold that releases fungal spores into the environment. It is endemic to the Ohio and Mississippi River valleys, the Great Lakes areas, and around the Saint Lawrence River. Pulmonary blastomycosis is caused by breathing in fungal spores from the air, especially after activities that disturb the soil. Most people who breathe in these spores, however, do not progress to symptomatic infection. Clinically apparent infection is more common in immunocompromised patients, but *Blastomyces* can also cause disease in immunocompetent individuals. Symptoms of blastomycosis can appear between 3 weeks to 3 months after exposure. Symptoms include fever, cough, night sweats, muscle aches, weight loss, chest pain, and fatigue. In some patients, blastomycosis can spread from the lungs and cause a disseminated infection affecting the skin, bones, joints, and the central nervous system. Animals, especially dogs, are also susceptible to blastomycosis, and symptoms are very similar to those of humans. Infection is not transmitted person-to-person or animal-to-person. Primary cutaneous blastomycosis is rare but can be a result of traumatic inoculation. The characteristic cutaneous finding is a verrucous lesion with irregular borders.

Blastomycosis can be diagnosed by direct microscopic identification of broad-based budding yeasts in fungal or cytology smears or histopathological staining of the infected tissue. Calcofluor white staining is typically used for fungal smears. Yeast forms of *Blastomyces* are typically large (8–15 μm) with broad-based budding and double-contoured cell walls that are refractile (Fig. 3.21). In histopathological specimens, *Blastomyces* can be difficult to identify solely with a hematoxylin and eosin (H&E) stain. A granulomatous or pyogranulomatous tissue response is frequently seen, but this finding is nonspecific for blastomycosis. Gomori methenamine silver (GMS) and periodic acid-Schiff (PAS) stains are better than H&E for examining *Blastomyces* fungal elements in tissue. *Blastomyces* can also be isolated in culture from respiratory specimens, skin, or other infected tissue. In culture, at 25–30°C, *Blastomyces* will grow as the mold form typically within 1–4 weeks. *Blastomyces* cultures typically have a beige-white colored growth on the obverse (e.g., top) of the agar (Fig. 3.22). A tape preparation of the colonies shows septate hyphae with unbranched conidiophores and single, terminal conidia that resemble lollipops (Fig. 3.23). At higher temperatures of 37°C, the colonies can convert from the mycelial phase to the yeast phase on media such as inhibitory mold agar.

Urine or serum antigen tests or serum antibody tests are also helpful in the diagnosis of blastomycosis. Immunodiffusion assays that measure antibody response to

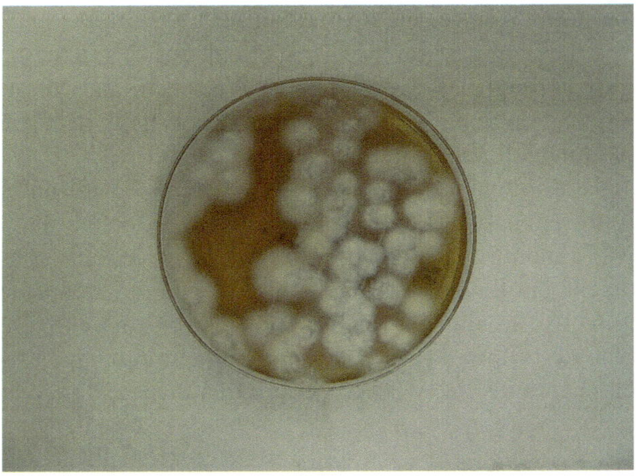

Fig. 3.22 Macroscopic culture of *Blastomyces* on inhibitory mold agar (IMA) plate (obverse view) with beige-white colored growth after 12 days

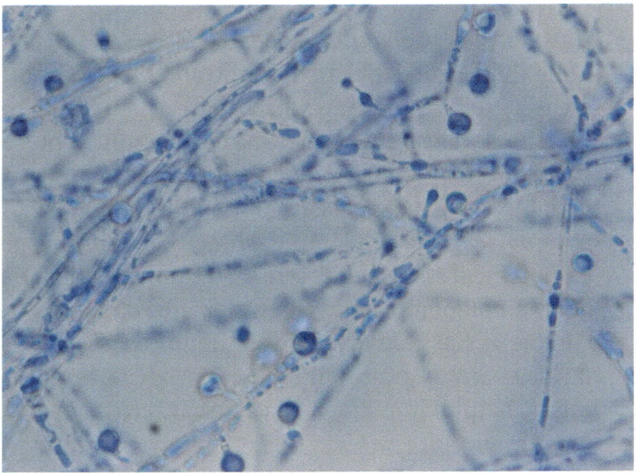

Fig. 3.23 Microscopic view (400× magnification) of tape preparation (lactophenol cotton blue stain) of the colonies shows septate hyphae with unbranched conidiophores and single, terminal conidia that resemble lollipops

Blastomyces antigen are relatively specific with moderate sensitivity. Enzyme immunoassays have greater sensitivity but lower specificity as compared to immunodiffusion assays. Thus, most laboratories initially perform an enzyme immunoassay which is then reflexed to immunodiffusion if positive. Complement fixation tests lack both sensitivity and specificity. Antigen detection in serum and urine is reliably used for the diagnosis of blastomycosis with greater sensitivity in urine than in serum. All antigen detection tests, however, have a high degree of cross-reactivity with other thermally dimorphic pathogens which cause endemic mycoses, most notably *Histoplasma*, but also *Paracoccidioides* species and *Talaromyces* (formerly known as *Penicillium*) *marneffei*.

Molecular identification of *Blastomyces* is also used for diagnosis. Real-time PCR assays can detect *Blastomyces* in various clinical specimens and even in paraffin-embedded tissue blocks. However, PCR can lack sensitivity when the organism burden is relatively low. Sequencing of the ribosomal genes, internal transcribed spacer (ITS) regions, and other repetitive sequences can differentiate *B. dermatitidis* from *B. gilchristii.* Both *B. dermatitidis* and *B. gilchristii* appear to have overlapping geographic distributions but *B. gilchristii* is limited to Canada and a few northern US states such as Wisconsin and Minnesota. Differentiation by the clinical microbiology laboratory is not usually necessary as the disease process and treatment do not differ between species.

Blastomycosis, when severe, is treated with amphotericin B, which is given intravenously. The azole agents are effective but must be administered for 6–12 months and sometimes even longer. Itraconazole is the most commonly used agent, but fluconazole, voriconazole, and posaconazole have also been shown to be effective treatments for blastomycosis. Urine antigen tests and chest radiographs can be followed over time to monitor improvements with therapy. Antigen levels appear to decline with successful treatment and increase with recurrence of illness.

Summary Points

- *Blastomyces dermatitidis* and *Blastomyces gilchristii* are thermally dimorphic fungi that are endemic to the Ohio and Mississippi River valleys, the Great Lakes areas, and the Saint Lawrence River.
- Pulmonary blastomycosis is the most common presentation, but cutaneous infections due to traumatic inoculation are also seen. Clinical presentation can be variable and can be catastrophic if left untreated.
- *Blastomyces* can be identified directly in clinical specimens by smears and histopathological stains by its size (8–15 μm), characteristic broad-based budding and double-contoured cell walls which are refractile.
- Culture is the gold standard for definitive diagnosis of blastomycosis. PCR can lack sensitivity especially when the organism load is low, and serological testing is helpful for diagnosis but can cross-react with other thermally dimorphic fungi.

- Amphotericin B is typically used to treat severe infections; azoles are second-line therapies. Cases of mild to moderate disease may be managed with azoles alone.

Selected References

Babady NE, Buckwalter SP, Hall L, et al. Detection of *Blastomyces dermatitidis* and *Histoplasma capsulatum* from culture isolates and clinical specimens by use of real-time PCR. J Clin Microbiol. 2011;49:3204.

Bariola JR, Hage CA, Durkin M, et al. Detection of *Blastomyces dermatitidis* antigen in patients with newly diagnosed blastomycosis. Diagn Microbiol Infect Dis. 2011;69:187.

Frost HM, Novicki TJ. *Blastomyces* antigen detection for diagnosis and management of Blastomycosis. J Clin Microbiol. 2015;53:11.

Jiang Y, Dukik K, Munoz JF, Sigler L, Schwartz IS, Govender NP, Kenyon C, Feng P, Gerrits Van den Ende B, Stielow JB, Stchigel AM, Lu H, Hoog S. Phylogeny, ecology and taxonomy of systemic pathogens and their relatives in *Ajellomycetaceae (Onygenales): Blastomyces, Emergomyces, Emmonsia, Emmonsiellopsis*. Fungal Divers. 2018;90:245.

Martynowicz MA, Prakash UB. Pulmonary blastomycosis: an appraisal of diagnostic techniques. Chest. 2002;121:768.

Saccente M, Woods GL. Clinical and laboratory update on blastomycosis. Clin Microbiol Rev. 2010;23:367–81.

Wheat LJ. Antigen detection, serology, and molecular diagnosis of invasive mycoses in the immunocompromised host. Transpl Infect Dis. 2006;8:128.

Case 7

Anisha Misra and Audrey N. Schuetz

A 72-year-old male with a past medical history significant for anaplastic large cell lymphoma and multi-organ metastases presented with fever of 39°C, nonproductive cough, and hypoxemia. He also had a diffuse rash with superficial ulcers which were thought to be paraneoplastic in origin. His current medications included brentuximab (a monoclonal antibody treatment for lymphoma), and he had recently received a 2-week course of prednisone 1 month before presentation. The patient was admitted to the hospital for further evaluation. A chest radiograph showed new areas of consolidation in the periphery of both lungs and diffuse interstitial pulmonary opacities. One month prior, he had a polymicrobial bloodstream infection with methicillin-resistant *Staphylococcus aureus*; the bacteremia cleared with treatment. Empiric therapy during this current hospitalization included cefepime and vancomycin. He decompensated overnight with worsening hypoxemia, difficulty breathing, and hypotension necessitating transfer to the intensive care unit where he underwent endotracheal intubation and was started on vasopressors. A complete blood count test indicated low hemoglobin of 6.8 g/dL (normal range 13.2–16.6 g/dL), hematocrit 20.4% (38.3–48.6%), an elevated platelet count of 539×10^9/L ($135–317 \times 10^9$/L), and a high leukocyte count of 23.9×10^9/L ($3.4–9.6 \times 10^9$/L). A basic metabolic panel showed elevated potassium 5.9 mmol/L (3.6–5.2 mmol/L), creatinine 1.87 mg/dL (0.74–1.35 mg/dL), and glucose 200 mg/dL (70–140 mg/dL).

The differential diagnosis included hospital- or community-acquired bacterial pneumonia due to *Streptococcus pneumoniae, Haemophilus influenzae, Chlamydia pneumoniae, Mycoplasmoides* (formerly *Mycoplasma*) *pneumoniae*, and *Legionella pneumophila*. Viral illnesses such as coronavirus disease 2019 (COVID-19) and atypical pneumonia with thermally dimorphic fungal pathogens were also high on the differential. The use of brentuximab has been associated with reactivation of cytomegalovirus (CMV) and increased risk of *Pneumocystis jirovecii* pneumonia; therefore, these entities were also considered in the differential. Initial blood cultures collected on admission remained negative after 48 h; his overall clinical picture was not consistent with bloodstream infection or endocarditis. CMV PCR and serum (1,3)-beta-D-glucan (BDG) were ordered. Bronchoalveolar lavage fluid (BAL) was sent for bacterial, fungal, and mycobacterial cultures, *Pneumocystis* smear and PCR, and severe acute respiratory syndrome-coronavirus-2 (SARS-CoV-2) PCR. The BDG level was 500 pg/mL (reference range <60 pg/mL). The *Pneumocystis* smear by calcofluor white stain is shown in Fig. 3.24; *Pneumocystis* PCR was also positive. The clinical team started intravenous (IV) trimethoprim-sulfamethoxazole (5 mg/kg) every 8 h. However, the patient continued to decompensate and passed away.

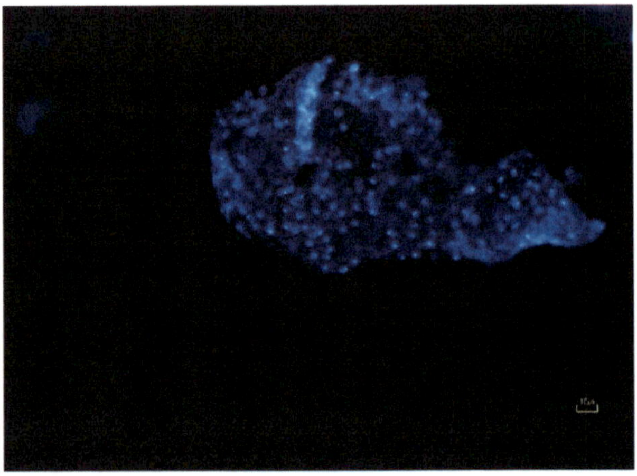

Fig. 3.24 Immunofluorescent stain of *Pneumocystis* cysts in bronchoalveolar lavage fluid (200× magnification). The bright dash-like structures resembling double parentheses and the lack of budding support the diagnosis of *Pneumocystis*

Discussion

Pneumocystis pneumonia (PJP) is a serious infection caused by the fungus *P. jirovecii*. The organism was initially known as *Pneumocystis carinii*, but the name was changed to *P. jirovecii* to distinguish it from the *P. carinii* species that infects rats. *Pneumocystis* infections are rare in healthy adults, and the organism colonizes the lungs of up to 20% of healthy adults. True infections occur predominantly in immunocompromised patients such as patients with HIV/AIDS, cancer, chronic lung disease, and those on chronic immunosuppressive therapy. *Pneumocystis* was first identified in patients with AIDS in the 1980s. Clusters of PJP were one of the first signs of the beginning of the HIV/AIDS epidemic. PJP was later classified as one of the AIDS-defining illnesses in the United States. It has undergone reclassification from a protozoan to a fungus based upon ribosomal RNA and gene sequence homologies.

The availability of antiretroviral therapy has reduced the incidence of PJP in patients with HIV; however, other immunocompromised patients remain at high risk. *Pneumocystis* spreads from person to person via aerosols. Symptoms can develop over several days or weeks and include fever, cough, dyspnea, chest pain, chills, and fatigue. Chest radiograph appearance can be variable, but the classic findings include diffuse, bilateral interstitial infiltrates. Diagnosis of *Pneumocystis* pneumonia relies upon visualization of the cyst or trophic forms in a lower respiratory tract specimen such as a BAL or bronchial washing. *Pneumocystis* cysts are round, 5–8 μm in size with thick walls and up to eight intracystic bodies, although the intracystic bodies are variably seen and are typically visible only on certain stain preparations. Trophic forms of *Pneumocystis* are pleomorphic, about 1–5 μm in size with a single nucleus. A *Pneumocystis* smear may be performed on induced sputum or BAL using the calcofluor white stain or an immunofluorescent stain consisting of fluorescein-labeled monoclonal antibodies. Lack of budding and presence of clearly visible intracystic bodies are helpful diagnostic features for *Pneumocystis* (Fig. 3.24). The immunofluorescent stain is more sensitive than calcofluor white and is the preferred diagnostic method for fresh specimens submitted to the microbiology laboratory. Diagnosis can also be obtained through histopathology of lung tissue or cytology of respiratory specimens. Organisms may appear in a honeycomb pattern, reflecting the alveolar outlines (Fig. 3.25). Gomori methenamine silver (GMS) stains the cyst forms, and Wright-Giemsa stains the trophic forms of *Pneumocystis*. On GMS, *Pneumocystis* cysts appear cup-like and may contain intracystic bodies which appear as double parentheses (Fig. 3.26). *Pneumocystis* cannot be cultured in the clinical microbiology laboratory. PCR is typically the first-line diagnostic test especially in less severely immunocompromised patients, since the organism load can be very low. However, qualitative PCR cannot differentiate between colonization and disease. A serum blood test to detect BDG, a carbohydrate cell wall component of many fungi, can also aid in a presumptive diagnosis of PJP when levels are greater than 60 pg/mL. However, the presence of BDG only provides a presumptive diagnosis of PJP, as elevated levels have been detected in patients with other fungal infections.

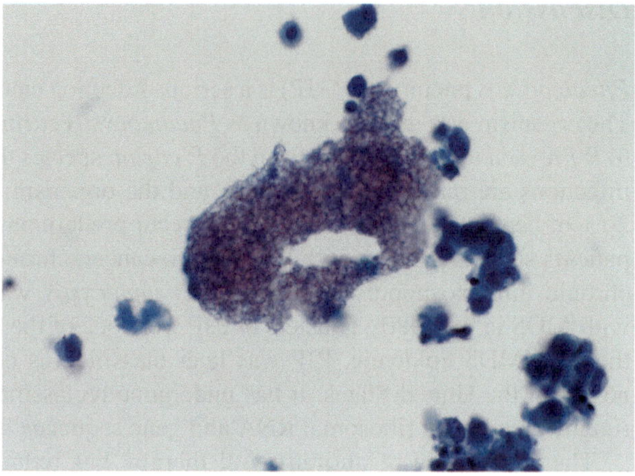

Fig. 3.25 Foamy exudates or honeycombing are common patterns of *Pneumocystis* pneumonia and are seen on Giemsa stain of this cytologic smear of bronchoalveolar lavage fluid from an HIV-positive patient with pulmonary infiltrates (400× magnification)

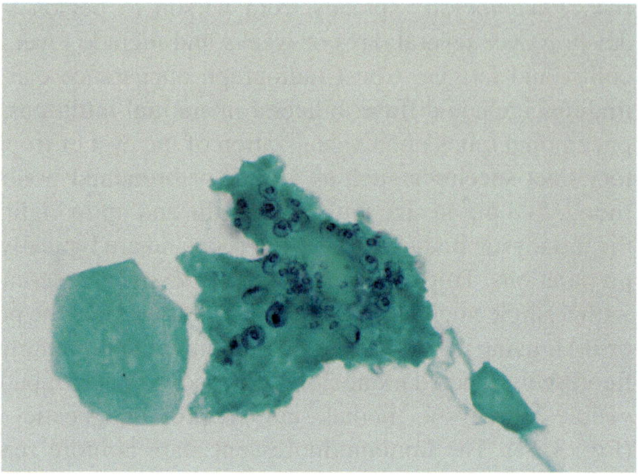

Fig. 3.26 GMS highlights the cyst forms of *Pneumocystis* on this bronchoalveolar lavage fluid of an HIV-positive patient with pulmonary infiltrates (1000× magnification). The double parentheses of the intracystic bodies are easily visualized

Currently, there are no vaccines to prevent *Pneumocystis* pneumonia. The most common prophylactic strategy is trimethoprim-sulfamethoxazole. Prophylaxis is often recommended for HIV patients with CD4 counts <200 cell/μL, stem cell transplant patients, certain solid organ-transplant recipients, and those on chronic corticosteroid therapy at a dose equivalent to at least 20 mg of prednisone for at least 4 weeks.

Summary Points

- *Pneumocystis* pneumonia should be considered in patients with an immunocompromised status, especially patients with HIV/AIDS, stem cell transplant patients, cancer patients, and patients taking long-term high-dose corticosteroids.

- Visualization of cystic or trophic forms in respiratory secretions or tissues can aid in diagnosis. Lack of budding and presence of intracystic bodies are key diagnostic features.
- *Pneumocystis* PCR is helpful in detecting the organism and is more sensitive than smears; however, a positive PCR cannot differentiate between colonization and disease.
- Presumptive diagnosis of *Pneumocystis* pneumonia can be made based on chest radiographs along with elevated serum (1,3)-beta-D-glucan (BDG) levels (>60 pg/mL); however, elevated BDG levels can be seen in other fungal infections.
- Trimethoprim-sulfamethoxazole is the most common prophylactic agent for *Pneumocystis* infections in immunocompromised individuals.

Selected References

Alanio A, Hauser PM, Lagrou K, et al. ECIL guidelines for the diagnosis of *Pneumocystis jirovecii* pneumonia in patients with haematological malignancies and stem cell transplant recipients. J Antimicrob Chemother. 2016;71:2386.

Bossart S, Mühlethaler K, Garzoni C, Furrer H. Is real time PCR preferable to the direct immunofluorescence in the diagnosis of *Pneumocystis jirovecii* pneumonia in HIV-infected patients? BMC Res Notes. 2020;13:235.

Doyle L, Vogel S, Procop GW. *Pneumocystis* PCR: it is time to make PCR the test of choice. Open Forum Infect Dis. 2017;4:193.

Kovacs JA, Gill VJ, Meshnick S, Masur H. New insights into transmission, diagnosis, and drug treatment of *Pneumocystis carinii* pneumonia. JAMA. 2001;286:2450.

Sax PE, Komarow L, Finkelman MA, et al. Blood (1→3)-beta-D-glucan as a diagnostic test for HIV-related *Pneumocystis jirovecii* pneumonia. Clin Infect Dis. 2011;53:197.

Case 8

Anisha Misra and Audrey N. Schuetz

A 75-year-old male with a medical history significant for ischemic cardiomyopathy, diabetes mellitus type 2, hypertension, and hyperlipidemia presented to the hospital emergency department with arrhythmias. He had had repeated hospital admissions for cardiac issues. During the hospital stay, it was noted that the patient had swelling and erythema of the right fifth digit and loss of the ability to flex his fingers (Fig. 3.27). The patient noted that the swelling had been progressive for the previous 3 months and initially started after he had fallen in his garden. The patient was

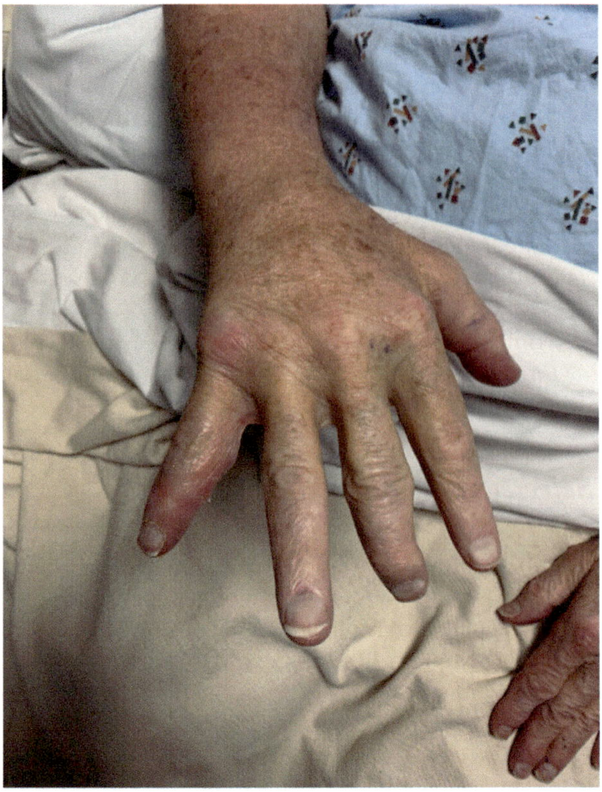

Fig. 3.27 Swelling of the right fifth digit with the swelling extending toward the wrist and elbows

afebrile without peripheral leukocytosis or systemic signs of infection. An ultrasound examination of his right upper extremity revealed findings consistent with inflammation of the distal right forearm and wrist, a complex fluid collection of the palm deep to the flexor tendon, and a possible developing abscess of the distal wrist. Aspiration of the right digit revealed 87,000 nucleated cells per high power field with 85% neutrophils, concerning for infectious tenosynovitis. An aspirate of the right digit was sent for bacterial, mycobacterial, and fungal cultures while awaiting surgical intervention and debridement. All cultures of the aspirate were negative for organism growth. Due to the chronicity of his symptoms and lack of acute systemic toxicity, an indolent infectious process was suspected. Differential included nocardiosis, cutaneous leishmaniasis, and mycobacterial infection including atypical mycobacteria—specifically *Mycobacterium marinum*. The patient underwent right hand and forearm incision and extensive debridement. Histopathologic examination of the debrided tissue showed extensive granulomas, necrosis, and both acute and chronic inflammation (Fig. 3.28). Operative tissue cultures grew a yeast form on inhibitory mold agar (IMA) plates incubated at 30°C. The yeast was sent for matrix-assisted laser desorption ionization–time-of-flight mass spectrometry (MALDI-TOF MS), but identification could not be made. Wet preparation of the colony showed rare, elongated yeast forms (i.e., cigar bodies). The isolated colony was re-processed

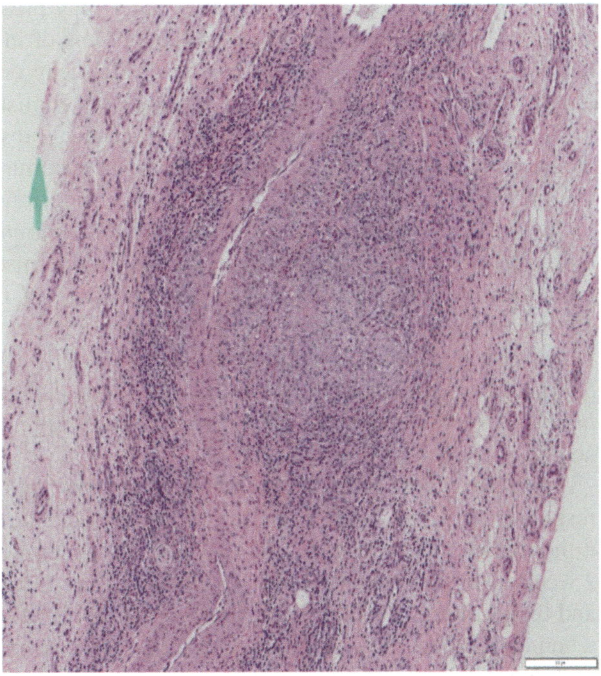

Fig. 3.28 Hematoxylin and eosin stain (100× magnification) of the forearm tissue with extensive granulomas, necrosis, and both acute and chronic inflammation

for MALDI-TOF MS with the tube-extraction method typically used for filamentous fungi, and subsequently resulted in a positive identification of *Sporothrix schenckii* complex. The mold form of *Sporothrix* grew after 10 days of incubation on IMA at 30°C, and tape preparations confirmed the diagnosis (Fig. 3.30). Treatment of choice is typically itraconazole, but the patient had several contraindications to azole therapy including significant cardiac disease, prolonged QT interval on electrocardiogram, and anticoagulation therapy with warfarin. Hence, he was treated with terbinafine. Due to worsening swelling a few months later, the patient was placed on long-term posaconazole therapy with adequate response.

Discussion

Sporothrix schenckii is a saprophytic mold that causes a cutaneous infection, sporotrichosis. *Sporothrix schenckii* resides on rose or barberry bushes, sphagnum moss, and in other mulches in its mycelial form producing infectious conidia. Due to its association with rose bushes, disease due to this agent is also called "rose gardeners' disease." Horticulturists, gardeners, farm laborers, and timber workers are most often infected through a small cut in the skin, resulting in lymphocutaneous infection. Unlike other thermally dimorphic fungi, *Sporothrix schenckii* only rarely causes pulmonary disease. Cutaneous sporotrichosis usually results in a small, painless bump that typically develops 1–12 weeks after exposure. The lesion is reddish-purple in appearance and eventually may present as an open ulcer that is very slow to heal. Systemic signs and symptoms of infection are typically absent, but bacterial superinfections at the site of ulcer can occur. Disseminated sporotrichosis is rare and mostly occurs in immunocompromised persons.

Diagnosis of sporotrichosis is based on clinical presentation and history of possible exposure to the fungus, alongside a positive culture. The characteristic cutaneous presentation can be similar in appearance to lesions caused by other pathogens including mycobacteria, *Nocardia*, *Francisella tularensis*, and *Leishmania brasiliensis*. Microbiologic diagnosis can be made using histopathological stains or culture from the active infection site. Microscopic and macroscopic analysis of the culture and MALDI-TOF MS can provide an identification of *S. schenckii* complex which can be further identified to the species level by gene sequencing. The *S. schenckii* complex includes *Sporothrix albicans*, *Sporothrix globosa*, *Sporothrix luriei*, *Sporothrix mexicana*, and the most commonly isolated agents are *S. schenckii* and *Sporothrix brasiliensis*.

Sporothrix schenckii can switch between the filamentous and yeast form in response to thermal stimuli and other environmental cues; hence, it is categorized as a thermally dimorphic mold. The organism can grow in culture within 2 days to 2 weeks. Macroscopically, colonies are smooth or wrinkled, white to creamy at first and then turning brown to black after a few days (Fig. 3.29). *Sporothrix schenckii*, when cultured at 25°C, demonstrates a filamentous form with thin, septate hyphae with conidia in small, clustered denticles (Fig. 3.30). The transition from mold to yeast form occurs at 35–37°C. A wet preparation of the colony from a plate incubated at 35–37°C, or demonstration of the organism on histopathologic examination of tissue, demonstrates yeast-like cells in various sizes and shapes. The yeasts are round to oval and 2- to 6-µm in diameter, and usually have elongated, cigar-shaped buds on narrow bases (Fig. 3.31). The fungus presents in human tissue as budding yeasts and cigar-shaped forms. Histopathological examination of the tissue with hematoxylin and eosin (H&E) stain can raise the differential of sporotrichosis in the

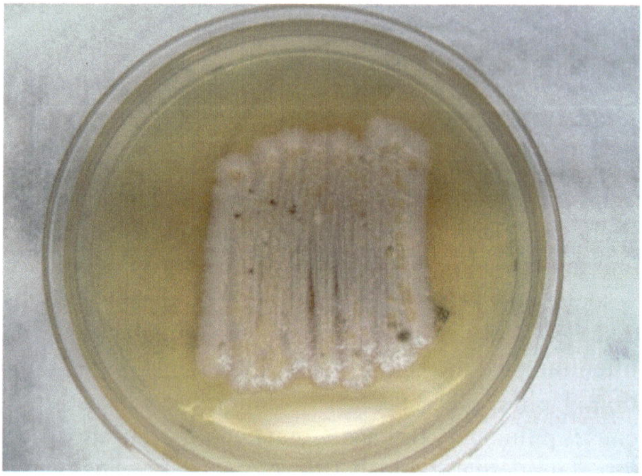

Fig. 3.29 Obverse view of *S. schenckii* cultured at 25°C on inhibitory mold agar (IMA) with smooth or wrinkled, white to creamy colonies

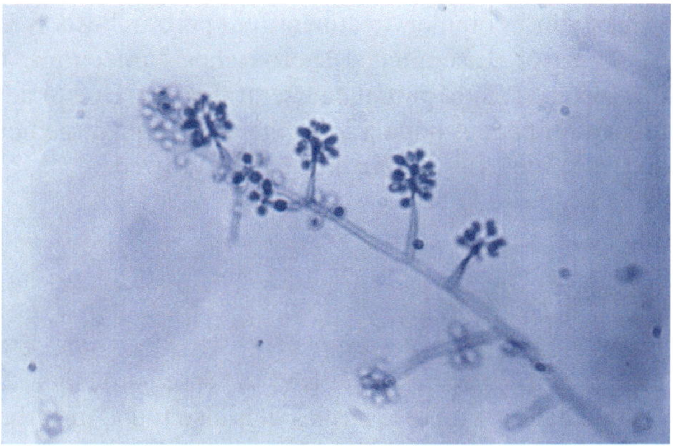

Fig. 3.30 Microscopic tape preparation (lactophenol blue stain, 970× magnification) of *S. schenckii*. Note the conidia in small, clustered denticles. Image courtesy of CDC Public Health Image Library

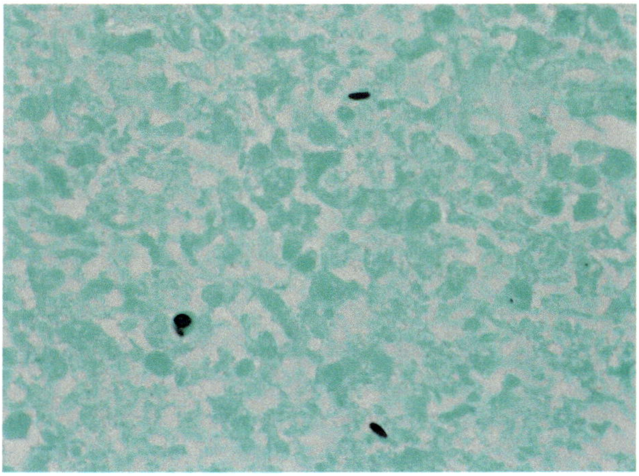

Fig. 3.31 Gomori methenamine silver (GMS) stain of forearm tissue demonstrates elongated yeast forms also known as cigar bodies and round to oval (1000× magnification)

correct clinical context, but other special stains such as Gomori methenamine silver (GMS) or periodic acid-Schiff (PAS) should be employed to visualize the yeast-like cells. Sporotrichosis usually elicits a pyogranulomatous inflammatory response. The forms can be rare in number and difficult to visualize on histopathology with H&E and even with GMS or PAS stains. Culture is the gold standard, but in cases that are culture-negative, serological testing using an enzyme-linked immunosorbent assay can be utilized. Serological tests are specific, but their sensitivity varies based on clinical presentation and are most helpful in the diagnosis of extracutaneous *Sporothrix* infections. Antigen tests for *Histoplasma* and *Blastomyces* can cross-react with *Sporothrix*. Molecular methodologies such as nested PCR or sequencing of the internal transcriber space (ITS) regions of the ribosomal RNA gene (rRNA) show high sensitivity and can differentiate among species within the *S. schenckii* complex, but they are not routinely used since identification to the complex level is adequate for diagnostic purposes and for therapy.

The most common treatment for sporotrichosis is itraconazole for 3–6 months. In severe or disseminated sporotrichosis infections, intravenous amphotericin B may be used as the initial course of therapy. Due to adverse effects of itraconazole on certain patient populations, other antifungal medications such as posaconazole and, in some cases, terbinafine can be utilized.

Summary Points

- *Sporothrix schenckii* can lead to "rose gardeners' disease," so named due to its association with rose bushes. It is also found in barberry and sphagnum moss and causes a cutaneous infection common in patients with exposure to horticulture.
- On wet preparation of yeast-like colonies, small, oval yeast forms 2- to 6-μm in diameter are seen. Elongated, cigar-shaped buds on narrow bases may also be seen.
- Culture of the fungus with identification by either MALDI-TOF MS or gene-specific PCR analysis is the diagnostic gold standard.
- Itraconazole is the first line of therapy followed by posaconazole.

Selected References

Barros MB, de Almeida Paes R, Schubach AO. *Sporothrix schenckii* and sporotrichosis. Clin Microbiol Rev. 2011;24:633–54.

Mahajan VK. Sporotrichosis: an overview and therapeutic options. Dermatol Res Pract. 2014;2014:272376.

Oliveira MM, Santos C, Sampaio P, et al. Development and optimization of a new MALDI-TOF protocol for identification of the *Sporothrix* species complex. Res Microbiol. 2015;166:102.

Rudramurthy SM, Chakrabarti A. Sporotrichosis: update on diagnostic techniques. Curr Fungal Infect Rep. 2017;11:134.

Chapter 4
Parasitology Cases

Marc Roger Couturier, Cameron A. Brown, Angela Ma, Santiago Delgado Fernandez, Audrey N. Schuetz, and James J. Dunn

Supplementary Information The online version contains supplementary material available at https://doi.org/10.1007/978-3-031-65092-5_4.

M. R. Couturier (✉)
Department of Pathology, University of Utah/ARUP Laboratories, Salt Lake City, UT, USA
e-mail: Marc.Couturier@aruplab.com

C. A. Brown
Departments of Pathology & Immunology and Pediatrics, Baylor College of Medicine, Houston, TX, USA
e-mail: Cameron.Brown@bcm.edu

A. Ma
Public Health Ontario, Department of Laboratory Medicine and Pathobiology, University of Toronto, Toronto, ON, Canada
e-mail: Angela.Ma@oahpp.ca

S. D. Fernandez
Caracas, Venezuela

A. N. Schuetz
Department of Laboratory Medicine and Pathology, Mayo Clinic, Rochester, MN, USA
e-mail: Schuetz.Audrey@mayo.edu

J. J. Dunn
Department of Pathology, Texas Children's Hospital, Houston, TX, USA

Department of Pathology & Immunology, Baylor College of Medicine, Houston, TX, USA
e-mail: jjdunn@texaschildrens.org

J. J. Dunn et al., *Challenging Cases in Diagnostic Clinical Microbiology*, https://doi.org/10.1007/978-3-031-65092-5_4

Overview of Diagnostic Parasitology Methods

Examination for parasites in human specimens is a relatively archaic procedure that has evolved over centuries, thanks to advances in technology such as microscopes, special contrasting stains, and non-microscopic detection methods such as antibody, antigen, and nucleic acid detection methods. Macroscopic parasites have been documented in ancient writings from societies in Egypt, Mesopotamia, Greece, Rome, and beyond. Diseases caused by microscopic parasites such as malaria, trypanosomiasis, and leishmaniasis have also been described in ancient writings though no mechanism for visualizing these organisms was available at the time. The early advent of stool examination for parasites is attributed to Antonie van Leeuwenhoek in the seventeenth and eighteenth centuries, who was the first person to detect and describe protozoa from fresh water, animal feces, and eventually his own feces using his early microscopes. His exploratory science was the origin of the modern-day wet-mount stool examination. British naval doctors adopted this method and have well-described records of detecting protozoa and helminths/helminth eggs in the stool of symptomatic sailors in the South Pacific during naval explorations in the 1700s and 1800s.

Parasites can be broadly classified in three major conceptual categories: protozoa (single-celled organisms including amoeba, flagellates, ciliates, stramenopiles, and apicomplexans), helminths (worms including nematodes, cestodes, trematodes, acanthocephalans), and macroscopic ectoparasites (arthropods of medical importance [e.g., fleas, ticks, mites, maggots] and leeches). The methods and approaches employed to detect and identify these categories are as diverse as the organisms themselves. Likewise, the breadth of specimens that may need to be examined is also quite broad.

Specimens

The most commonly submitted specimen for parasite examination is stool. Stool is easily collected and can be a used as a specimen to detect over 20 different protozoa and helminths. Though most stool specimens will be collected from patients with diarrhea, it is also acceptable to test formed (*i.e.*, non liquid) stool, as many parasitic infections do not cause diarrhea. Whole blood is another commonly interrogated specimen type and is the primary specimen when *Plasmodium*, *Babesia, Trypanosoma,* and filarial nematodes are in the differential diagnosis. Serum is also commonly collected for antibody detection for a number of parasites. All other specimen types collected for parasite detection and examination have a much more limited differential, typically just one or two organisms. These specimens can include, but are not limited to, cerebral spinal fluid (CSF), respiratory specimens (sputum, bronchial alveolar lavage, tracheal aspirate), cyst or abscess fluid/aspirate, tissue biopsy, skin biopsy or scapings, ocular specimens (corneal scrapings, vitreous fluid), duodenal or colonic aspirates, and lymph node aspirates.

Stool Exam Techniques

Fresh Smear Preparation Stool examinations can be performed using a number of different stains and techniques. The original stool examination described by van Leeuwenhoek is the fresh wet-mount exam. In this examination, fresh (unfixed, unpreserved) stool is added to a microscope slide with saline and a coverslip is applied, then the specimen is visualized on the microscope. Formed stools are generally not used for this exam. Motile trophozoites can be detected with the saline prep. A second preparation using iodine is performed using another small sample of the stool. In this preparation, the organism is no longer motile but the iodine provides contrast to better detect organisms and visualize defining structures. This examination requires that stool be freshly collected and tested, which makes it less practical for most modern clinical parasitology laboratories which are not located near the bedside. Note, this preparation technique can also be performed on freshly collected CSF specimens to look for motile trophozoites of free-living amoeba (e.g., *Naegleria* and less commonly *Acanthamoeba* and *Balamuthia*).

Ova and Parasite (O&P) Examination from Fixed Stool The conventional O&P exam consists of a permanently stained trichrome slide and a concentrated wet-mount preparation with or without iodine. The trichrome stain was traditionally prepared from unconcentrated stool; however, recent studies have shown that concentrated stool using vertical filtration/centrifugation can also yield equivalent sensitivity with just a single preparation. Stool concentration can also be achieved by passive sedimentation or chemical concentration methods. Regardless of the manner of concentration, this step is key for maximum sensitivity for multiple stool examination techniques. The trichrome smear is prepared from fixed stool and is subsequently dried, stained, and visualized under 1000× magnification to detect trophozoites and cysts of protozoa. The cytoplasm of protozoan trophozoites will stain blue-green, with occasional purple features. Cysts tend to be slightly more purple or faint pink (especially *Entamoeba* spp.). Nuclei and inclusions (chromatoidal bodies, red blood cells (RBCs), bacteria, and Charcot-Leyden crystals) stain red (Fig. 4.1). The concentrated wet mount is prepared by adding a drop of concentrated stool to a microscope slide and mixing it with a drop of iodine followed by a coverslip. These preparations are viewed at 100× and 400×. The concentrated wet mount is used to detect nematode larvae, helminth eggs, and protozoal cysts and trophozoites (Fig. 4.2).

O&P examinations are not overly sensitive for diagnosis of intestinal parasitic infections. A single O&P examination is only 60% sensitive for protozoal infection. Three unique stool specimens collected over 3–5 days are recommended for optimal sensitivity. Even with three collections, sensitivity only approaches 90%. In the case of some helminths that only produce eggs with irregular frequency, the number of specimens may need to be increased to 5–7 unique specimens over 7–10 days.

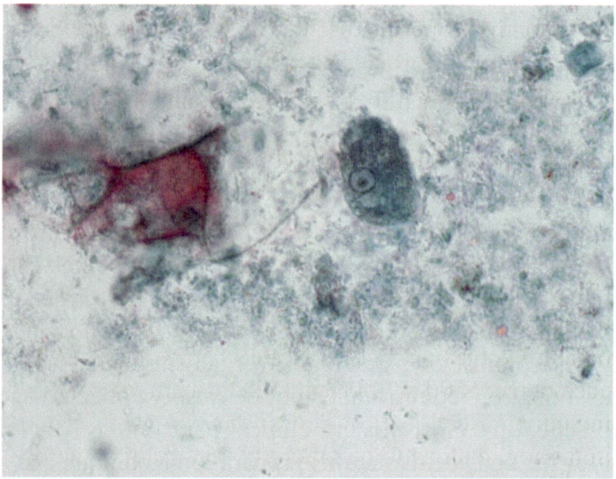

Fig. 4.1 Trichrome stain from concentrated stool showing *Entamoeba coli*. Green, blue/purple, and red staining is visible

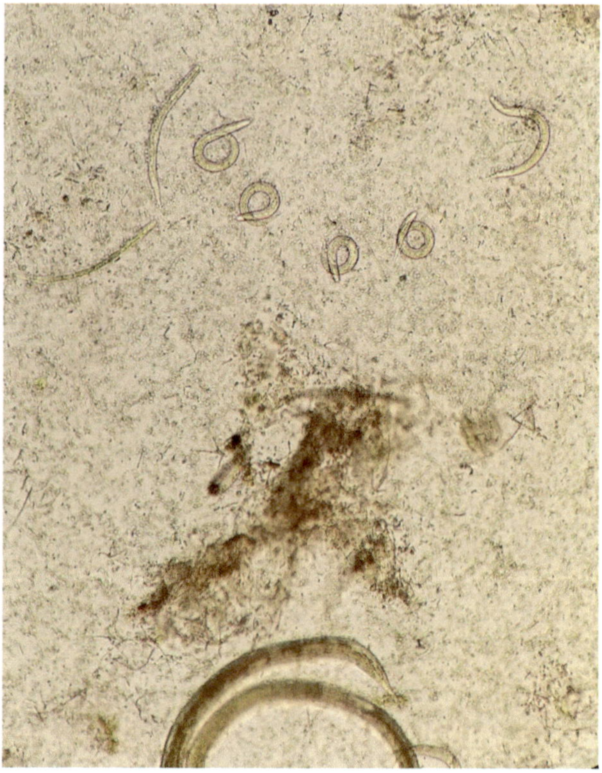

Fig. 4.2 Wet-mount iodine preparation of concentrated stool showing adult female *Strongyloides* (lower part of image) and numerous L1 larvae (upper part of image) (400× magnification)

Stool Stains for **Cryptosporidium** *and the Coccidia*

Cryptosporidium and intestinal coccidia (*Cyclospora* and *Cystoisospora*) cannot be readily detected by trichrome stain as the oocysts do not retain this stain. *Cryptosporidium* measures 4–6 μm, *Cyclospora* measures 8–10 μm, and *Cystoisospora* measures 20–33 μm. *Cystoisospora* can be detected in concentrated wet-mount preparations with relative ease due to its size; however, *Cyclospora* and *Cryptosporidium* are very difficult to identify from a standard bright-field wet-mount exam since they are small and can easily be confused for yeast, fat globules, and *Blastocystis*. Because of these limitations, special stains are typically used to improve detection of these organisms. The most commonly used stains are the modified acid-fast methods, either the modified Kinyoun or modified Ziehl-Neelsen stains from fixed, concentrated stool. For each method, one or two drops of stool are placed on a microscope slide and smeared to confluence. The smears should not be too thick, and should be roughly thin enough to read newsprint through the unfixed smear. Akin to the O&P, three unique specimens collected over 3–5 days is recommended to optimize sensitivity.

The modified Kinyoun is fixed in methanol and air-dried, whereas the modified Ziehl-Neelsen is heat-fixed and air-dried. The specimen is then stained using the appropriate staining technique and visualized at 1000×. The modified Kinyoun is stained with Kinyoun's carbol fuchsin and decolorized with 1% sulfuric acid and counterstained with methylene blue or malachite green (Fig. 4.3). This version of the acid-fast stain is more popular because it does not require a heating step and uses a mild acid decolorizer. The modified Ziehl-Neelsen is stained with hot carbol fuchsin and decolorized with 5% sulfuric acid. Slides are counterstained with either methylene blue or malachite green (Fig. 4.4). With both stains, artifacts can be commonly seen, including yeast and fat globules, which can both stain dark red. For both methods of acid-fast stain, the target organisms will stain a pink/fuchsia color. All three organisms can occlude the fuchsin entirely, resulting in cleared or "ghost" forms. This phenomenon is seen more often with the coccidia than *Cryptosporidium*.

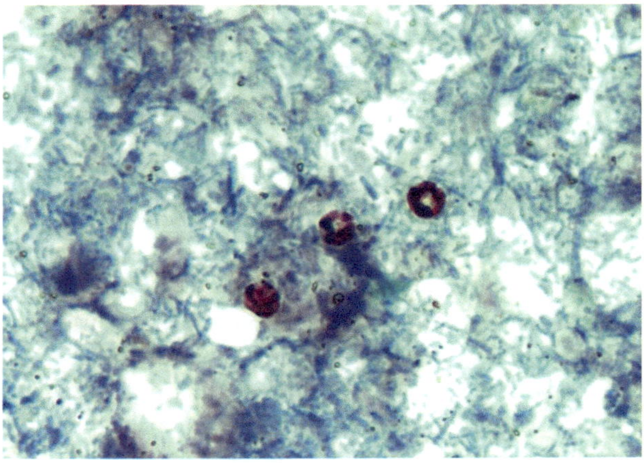

Fig. 4.3 *Cryptosporidium* spp. (4–6 mm) in concentrated stool stained with modified Kinyoun stain, counterstained with malachite green (1000× magnification)

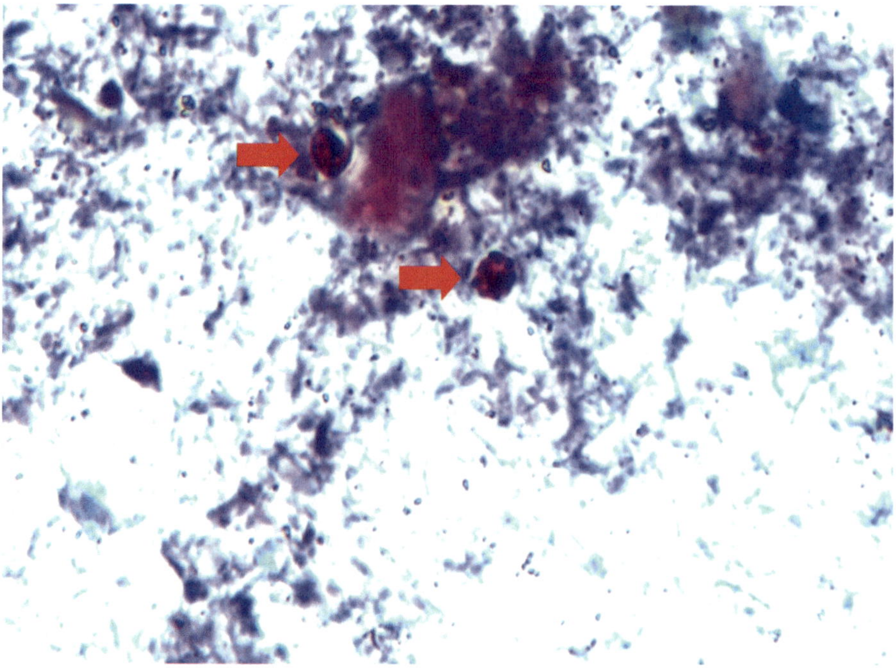

Fig. 4.4 *Cryptosporidium* spp. (4–6 mm, see arrows) in concentrated stool stained with modified Ziehl-Neelsen stain, counterstained with methylene blue (1000× magnification)

Another common stain for this group of organisms is the modified safranin stain. This stain is optimal for staining *Cyclospora* oocysts and results in significantly fewer ghost forms. This stain can also be useful in the identification of *Cryptosporidium* and *Cystoisospora*, albeit with no advantage over modified acid-fast for those organisms. The modified safranin stain requires the slide to be immersed in acidified safranin solution and brought to a boil using either a hot plate or microwave for 1 min to heat. The slide is counterstained with methylene blue and visualized at 1000× (Fig. 4.5). Due to the heating step and relatively messy procedure, few laboratories use this method, despite its advantage for *Cyclospora* detection.

UV Autofluorescent Microscopy

The intestinal coccidia, *Cyclospora, Cystoisospora,* and *Sarcocystis,* all have oocysts that autofluoresce under broad range ultraviolet (UV) light without the need for any stains or other modification. This method is fast, easy to prepare and screen, is inexpensive, and more sensitive than the permanent stains listed in the section above. Many helminth eggs also display autofluorescence under UV illumination, which is an added benefit to this method.

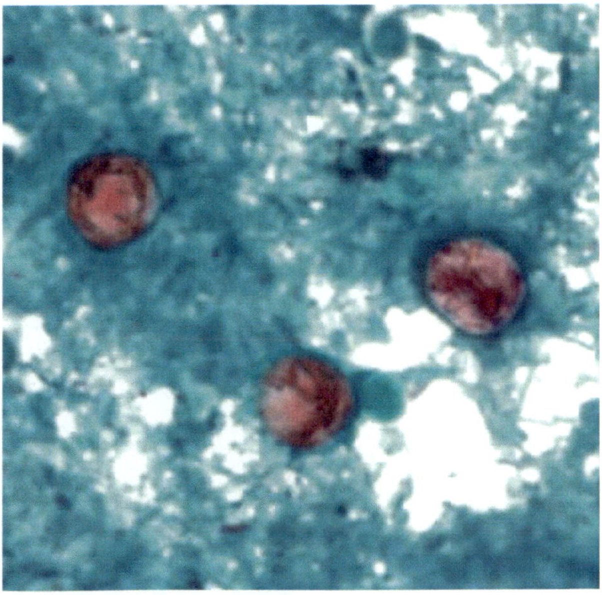

Fig. 4.5 *Cyclospora* (8–10 mm) in concentrated stool stained with modified safranin. (Image courtesy of the CDC's DPDx. https://www.cdc.gov/dpdx/cyclosporiasis/index.html)

A standard preparation should be prepared like the O&P concentrated wet mount described above. It is critical that iodine never be added to these preparations as it will quench the fluorescence. Saline should also not be added unless the specimen is too thick or mucoid to prepare an adequate quality wet mount. If a specimen has very dense fecal debris, it may be advisable to prepare two preparations on the same slide under two individual coverslips. One specimen can be prepared from the stool only and the other can be thinned with a drop of saline. Specimens should be visualized at 200–400× magnification. Oocyst walls will appear blue with an excitation filter of 330–365 nm (Fig. 4.6). A blue excitation filter of 450–490 nm will generate a less-intense green color. Oocysts observed with a barrier filter at 510–530 nm will appear green. Care should be taken to minimize the amount of UV light exposed to a single field of view. The specimen will be photobleached if illuminated too long. The field of view will typically fluoresce again if allowed several minutes without illumination. If a field of view has suspicious structure(s) that require additional confirmation by another lab member, the shutter should be applied until viewed.

Macroscopic Examination of Helminths

Arthropods and helminths are frequently submitted to diagnostic laboratories for identification. These organisms can be fragments or intact organisms depending on the nature of collection/removal from the host. Helminths may be recovered from a variety of clinical specimens and, in general, the specimen source should

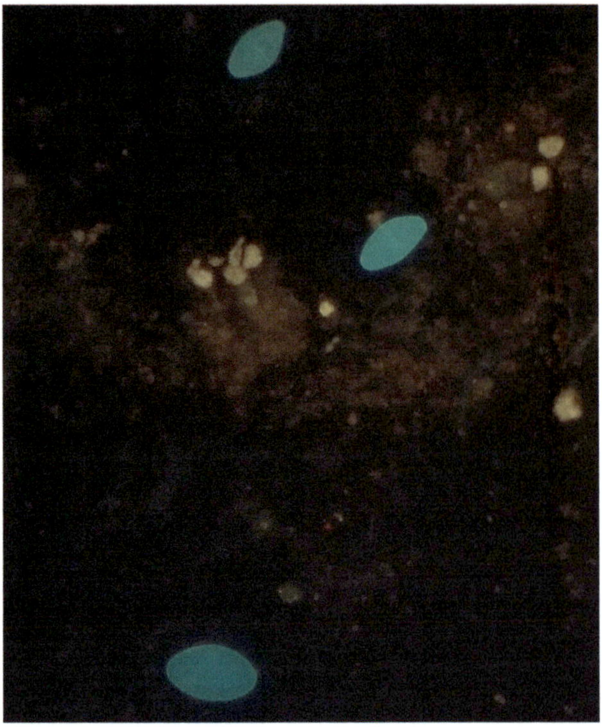

Fig. 4.6 UV autofluorescence of *Cystoisospora* in concentrated stool in a wet-mount preparation. A 375 nm excitation filter was used, generating a blue fluorescence (400× magnification)

be noted, as it can be helpful in correlating an identification to a causative infection. Common sources of gross organisms include: stool, skin, ocular sources, lungs, biliary tree, and aspirates of various sites (e.g., colon, stomach). The most frequently encountered parasites from enteric sources include nematodes (*Ascaris lumbricoides*, *Trichuris trichiura,* anasakids, and *Enterobius vermicularis*) and cestodes (*Taenia* spp., fish tapeworms, and *Dipylidium caninum*). Ocular-derived specimens are typically either *Loa loa* or *Dirofilaria*. All these organisms can be identified simply based on their gross morphologic features (length, width, features of mouthparts, external tegument features, tail structure, etc.). When examining a worm macroscopically, it is imperative that the worm be laid in a petri dish with fixative (alcohol or formalin) to ensure the specimen does not desiccate (Fig. 4.7).

In some cases, it may be necessary to view internal features such as the esophagus, intestine, ovaries, and eggs *in utero*. This is best achieved by clearing the helminth tissue in lactophenol and viewing under a dissecting scope (Fig. 4.8). Eggs can also be expressed from an adult gravid female nematode by gently pressing on the posterior end of the worm against a glass slide. In the case of cestodes, gentle pressure can be applied, or a rolling motion can be made, to express the eggs from the genital pore. After expressing any eggs, the eggs can be viewed with a drop of saline or iodine on the slide, and an identification should be achievable. It is important to consider that when using this

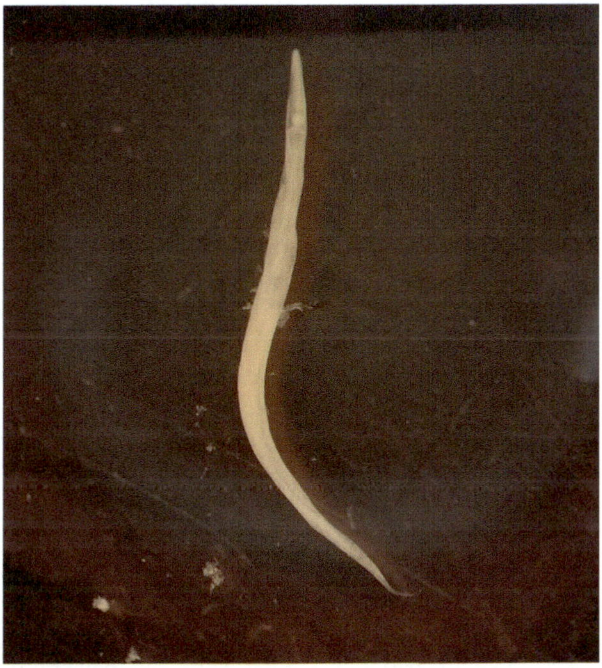

Fig. 4.7 Macroscopic examination of *Enterobius vermicularis* collected from a colonic wash. Specimen was visualized with a stereoscope

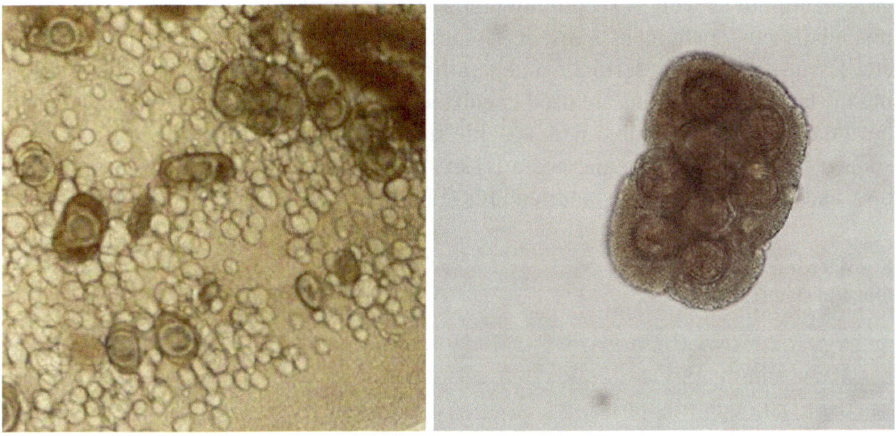

Fig. 4.8 *Dipylidium caninum* proglottid cleared in lactophenol with eggs *in utero* (left panel). Note the egg packets contain 1–5 eggs and are underdeveloped. The right panel shows an egg packet found in concentrated stool. This packet contains 9–10 eggs and is fully developed

method, some eggs may not be fully developed and thus may not match the textbook size and morphologic characteristics. One such example is *Dipylidium caninum*, for which egg packets will often only contain 2–4 eggs when expressed manually (or viewed through lactophenol cleared tissue), versus ~4–10 eggs when passed naturally (Fig. 4.8).

Macroscopic Examination of Arthropods

Medically important arthropods are usually recovered from skin or hair, but may also be found in and around human habitations. Examples of some of the medically important arthropods that a clinical laboratory may encounter include: ticks (adult, nymphal, and larval), lice and nits (eggs), bed bugs, fleas, mites, and myiasis-causing fly larvae (Fig. 4.9). Proper identification and reporting can be difficult for arthropods. Some of the common variables that affect the ability to identify an arthropod can include: the condition of specimen submitted, differentiation of clinically relevant organisms from incidental findings of non-parasitic organisms, and competency of staff members to identify organisms to a medically appropriate level. In general, arthropods should be identified to the level that is necessary for clinical care. Many laboratories will receive arthropods that are non-parasitic in nature, and these should not be identified at a high level of taxonomy, in an effort to avoid implying clinical significance (e.g., centipedes, moths, bees, wasps, ants).

Histopathologic Examination

Histopathologic examination and sectioning can be used in two different ways for identifying parasitic infections. This can include paraffin embedding and sectioning a gross helminth as well as sectioning human tissue suspected of having a parasite (or potentially other pathogens) present. Tissues can include biopsies, touch preparations, etc. Tissues are fixed in formalin and paraffin embedded for sectioning. After sectioning the target tissue, stains are used to enhance specific features. Some organisms will have unique features visible with specific stains. In general, most specimens will be stained with hematoxylin and eosin (H&E). The H&E can provide visualization of host and parasite structures and features (Fig. 4.10). H&E will stain most parasites,

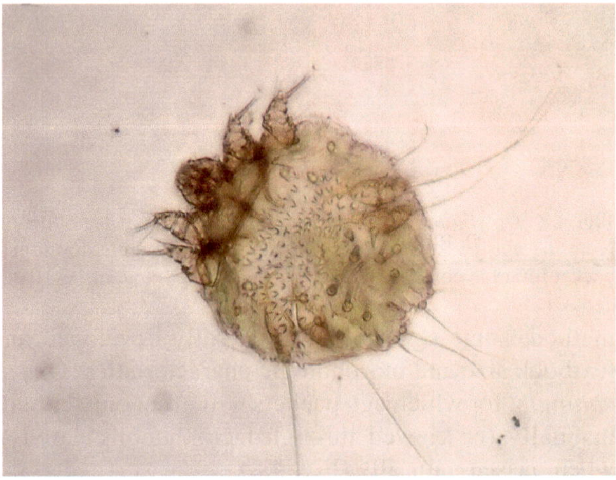

Fig. 4.9 *Sarcoptes scabiei* found in skin scrapings

but not all parasites show the same clarity with H&E, necessitating the use of other types of stains. One such example is the periodic acid-Schiff stain, which makes *Leishmania* and coccidia stand out very prominently against the host tissue (Fig. 4.11). Giemsa stain is also helpful for cutaneous *Leishmania* infections. Very few anatomic pathologists are trained to identify parasites to genus level in tissue. Expert consultation should be sought in most cases of suspected tissue parasites.

Parasite Culture

Free-Living Amoeba Culture Very few parasites are cultured in clinical laboratories. The most commonly cultured organisms are *Strongyloides* and free-living amoeba (FLA), in particular *Acanthamoeba*. In the case of the FLA, the organisms are actually propagated on solid media. The media is a non-nutritive agar that is

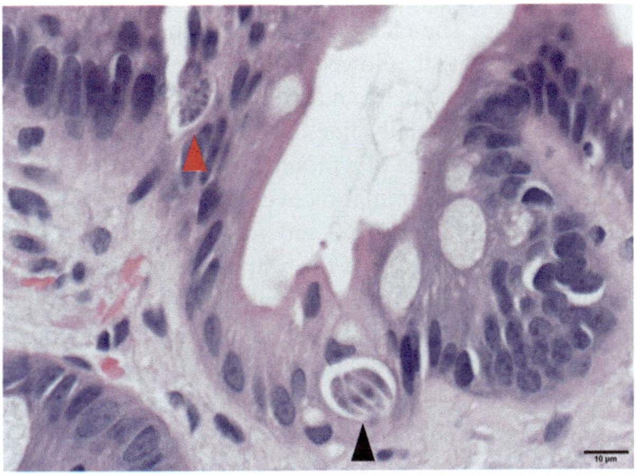

Fig. 4.10 Large bowel biopsy stained with hematoxylin and eosin showing merozoites (black arrowhead) and microgametes (red arrowhead) of *Cystoisospora*

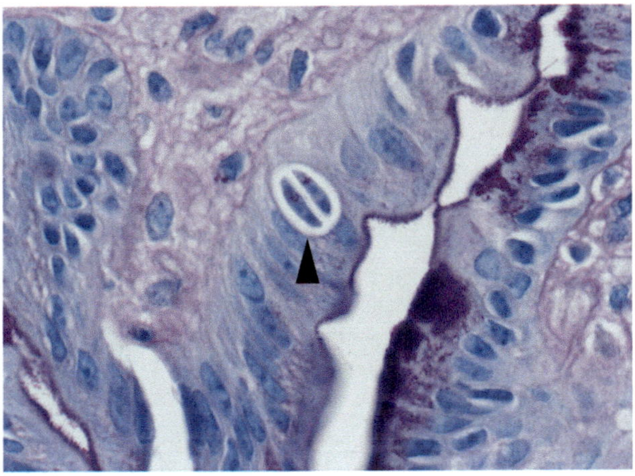

Fig. 4.11 Large bowel biopsy stained with periodic acid-Schiff showing merozoites (black arrowhead) of *Cystoisospora*

seeded with a bacterial lawn. This method is specifically used for *Acanthamoeba* and *Naegleria* since both organisms use bacteria as a primary food source. CSF can be inoculated in these cultures to detect cases of primary amoebic encephalitis or granulomatous amoebic encephalitis. Corneal scrapings and contact lens and/or solution can also be inoculated onto these media to recover *Acanthamoeba* in cases of suspected amoebic keratitis. Tissue culture cells of human/primate brain origin can also be used to propagate *Balamuthia* from CSF specimens from patients suspected of granulomatous amoebic encephalitis. *Balamuthia* will not grow on the non-nutritive agar with bacteria because it does not use bacteria as a food source. In nature, *Balamuthia* feeds on other single-celled organisms including other FLA. Trophozoites and cysts can be seen in FLA culture. Shortly after inoculation, trophozoites will be predominant, with eventual conversion to cysts as nutrients become depleted (Fig. 4.12).

Strongyloides **Agar Plate Culture** *Strongyloides* culture does not actually involve "culturing" larvae in the traditional sense, but rather visualizing the migration of motile larvae. The enteric flora from a patient's stool (or other non-sterile site with bacteria present) is allowed to grow on an agar plate and the *Strongyloides* larvae migrate across the plate and drag the bacteria with it resulting in "tracks" (Fig. 4.13). The tracks are actually lines of bacterial colonies spread across the surface, which indicate the presence of a motile larval nematode. The most recent culture conditions described for the optimal recovery of *Strongyloides* include nonselective media at a pH of 6.0 with 0.5% NaCl and yeast extract. The plates are inoculated in the center with approximately 2 g of fresh stool and incubated at 28–30 °C with the agar surface facing up. The plates must be taped shut to ensure no larvae are able to migrate out of the plate. Plates should be viewed at 2 days with an inverted microscope for any evidence of "tracks" and every day afterwards for 6 days before discarding the plates. If any tracks are seen, the plate should be moved to a biosafety cabinet and the tape carefully removed. The plate should be kept flat during this entire process to avoid any

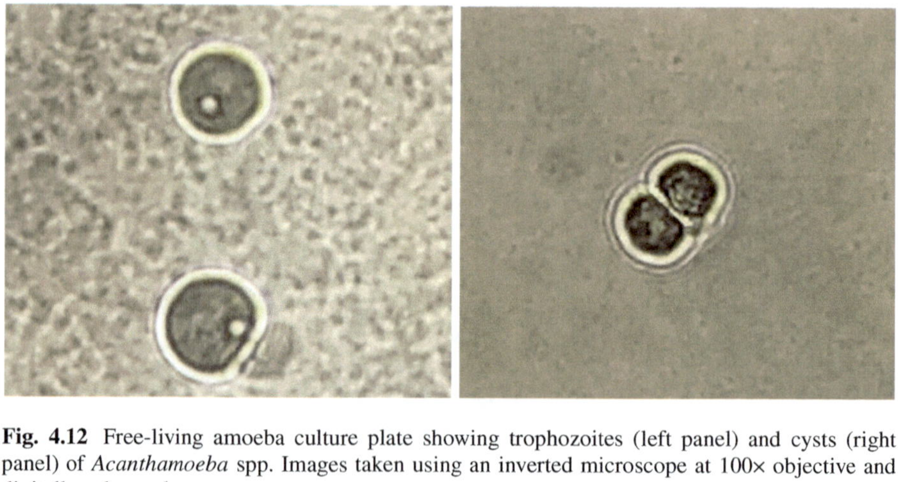

Fig. 4.12 Free-living amoeba culture plate showing trophozoites (left panel) and cysts (right panel) of *Acanthamoeba* spp. Images taken using an inverted microscope at 100× objective and digitally enhanced

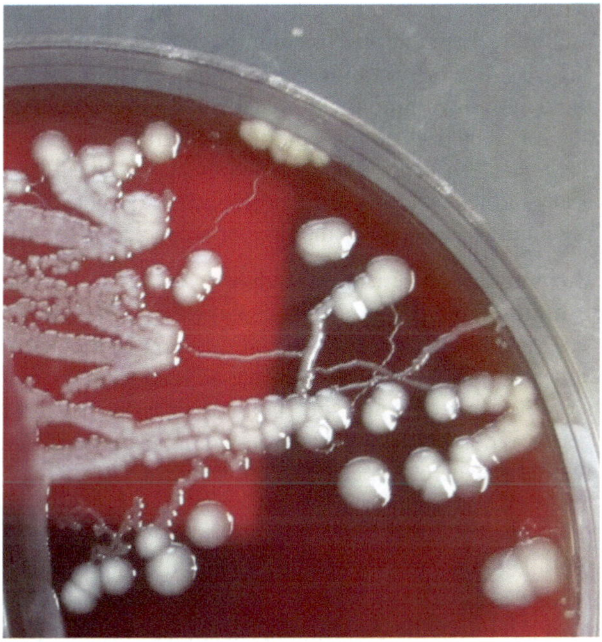

Fig. 4.13 Agar plate culture of stool with *Strongyloides* "tracks" visible as bacterial colonies

spillage of condensation, which could contain infectious larvae. Once the tape is removed, the surface of the plate should be flooded with 10 mL of 10% formalin and allowed to fix for 30 min. The formalin is then pipetted into a centrifuge tube and pelleted at 500 × g for 5 min. The sediment should then be sampled for a standard wet mount and screened at 100× or 200× magnification for the presence of larvae. If larvae are seen, 400× magnification should be used to identify the larvae.

The procedure is not without biosafety concerns, as there are live nematodes, which, in the case of *Strongyloides* or hookworm, can directly infect the host via skin penetration. Eggs of several nematodes can hatch on these plates and be seen in the subsequent concentrate. Extreme care must be taken when handling these plates and manipulating the plates for eventual microscopy.

Serology

Serology for parasites is challenging as many of these organisms elicit antibodies that are not specific to a single species or even genus. For instance, many nematodes will induce cross-reactive antibodies against other nematodes due to conserved structural proteins that are found in their tegument. Most parasites (especially helminths) have evolved cunning immune evasion mechanisms, either by decreasing the host immune response or through molecular mimicry of their antigens to resemble those found in the host. Extraintestinal protozoa are often detectable with serology with greater specificity than with helminths. Serology tests are available in multiple formulations including enzyme-linked immunosorbent assay based (ELISA) or chemiluminescent immunoassay (CIA), western or immunoblots, and rapid antibody immunochromatographic assays (ICA). The majority of these assays are for detection of IgG antibodies. With very rare exception (e.g., *Toxoplasma, Trypanosoma cruzi*), IgM is not used for detection of parasitic infections since many of these infections are prolonged in duration and IgM antibodies would not be predictably detected by the time clinical care is sought.

Enzyme-linked Immunosorbent Assay (ELISA) and Chemiluminescence Immunoassay (CIA) Most serology assays for parasites are ELISA- or CIA-based. The specificity and sensitivity of different ELISA/CIA methods are highly variable depending on several factors. The antigen source for an assay is one factor that can shift the sensitivity and specificity of an assay. For example, if a highly purified antigen or recombinant/synthetic antigen is used in the assay, it is likely that the assay will have excellent specificity but decreased sensitivity. This can be a valuable characteristic for a confirmatory assay or for ruling out potentially cross-reactive infections. Alternatively, a whole organism lysate-based or partial-purified antigen source will likely result in an assay that has excellent sensitivity but lower specificity. This type of assay design may be very good for use as a screening test that is confirmed by a more specific confirmatory test (see "blots" below). Alternatively, this type of design may be used for mass screening purposes of a specific population (e.g., immigrants, refugees), where it is important to detect all potential cases for empiric therapy (or "deworming" as it is sometimes referred). This is common for infections such as schistosomiasis and strongyloidiasis. Sensitivity can also be increased simply with different detection methods such as including a signal amplification step or a more sensitive optical measurement system (e.g., chemiluminescence, fluorescence).

Blots Western blots or immunoblots are another modality of serology that can be used to aid in the diagnosis of parasitic diseases. In general, blots are not used as standalone diagnostic tests and rather are endorsed as confirmatory assays following a "screening" ELISA. The advantage of a blot is that the specific antigen reactivity can be analyzed by viewing the banding patterns. This allows for a determination of specificity to be made. In an ELISA, the antibody reactivity is measured as an absorbance from the assay, without any information about the nature of the binding. In a blot, specific criteria or patterns can be established for interpretation. Alternatively, a nonspecific binding pattern can be easily identified by blot. In general, blots are not often used, though confirmatory blots for *Echinococcus* and *Taenia solium* are invaluable for the diagnosis of these organisms.

Antigen Detection

Antigen detection assays are available in two primary testing formulations: rapid diagnostic tests (RDTs) and ELISA-based tests. ELISAs are primarily used for stool testing for organisms such as *Giardia, Cryptosporidium,* and *Entamoeba histolytica*. These assays were introduced into clinical use approximately 20+ years ago and were valuable tools for laboratories that could not maintain traditional stool microscopy. Alternatively, they have been employed as a "screen out" for patients at high risk for said infections, with selective reflex to the more cumbersome microscopy methods. RDTs can also be used for stool parasite testing for the aforementioned organisms, including some multianalyte test cards/cartridges that detect *Giardia* and *Cryptosporidium* +/− *Entamoeba* spp. These assays are easier to use than the ELISA counterpart and can be more attractive for very low volume testing settings.

RDTs are also used for blood specimens. The most widely adopted RDTs are for malaria detection; however, RDTs for filariasis are also available in some countries. Malaria RDTs have become primary screening tests in endemic and non-endemic areas due to their low cost and ease of use. Blood is applied to a membrane directly or after a diluent step and allowed to migrate through the membrane. Antibodies that bind a *P. falciparum*-specific protein and/or pan-malarial protein are captured in the flow device and aggregate to make a visible line. Different assays have different targets and interpretations, so laboratories should follow the package insert carefully to ensure an accurate interpretation of the test is made. In general, these tests are quite sensitive and specific, but not as sensitive as a blood smear at very low parasitemias. As a result, it is recommended that an RDT be followed up with a blood smear regardless of whether the result is positive or negative.

Blood Smears

Blood smears are used to diagnose several parasites, including *Plasmodium*, *Babesia*, *Trypanosoma*, and the filarial nematodes (*Wuchereria*, *Brugia*, *Mansonella*, and *Loa loa*). Peripheral or finger-stick blood can be stained with Wright, Wright-Giemsa, or Giemsa stains. Giemsa stains are preferred for parasite identification, as some staining features are more pronounced for the parasite. Wright and Wright-Giemsa stains are also commonly used; however, some staining features may not be readily visible and/or are sometimes difficult to interpret due to different pH ranges of these stains. A complete blood smear exam includes a minimum of two thin blood smears and two thick blood films. The thick blood films contain higher density of blood than the thin films and are lysed, releasing the organisms from the infected RBCs (Fig. 4.14). This film allows for sensitive detection of parasites at the expense of losing some morphologic features. The thin smears are intended to allow for intracellular parasites to be visualized with all morphological features within the parasitized RBCs (Fig. 4.15). The percentage of infected RBCs can also be calculated easily with the thin smear. To calculate a parasitemia, 1000 red blood cells should be counted over approximately 20 unique fields of view. The total number of red blood cells parasitized is divided by the total number of red blood cells

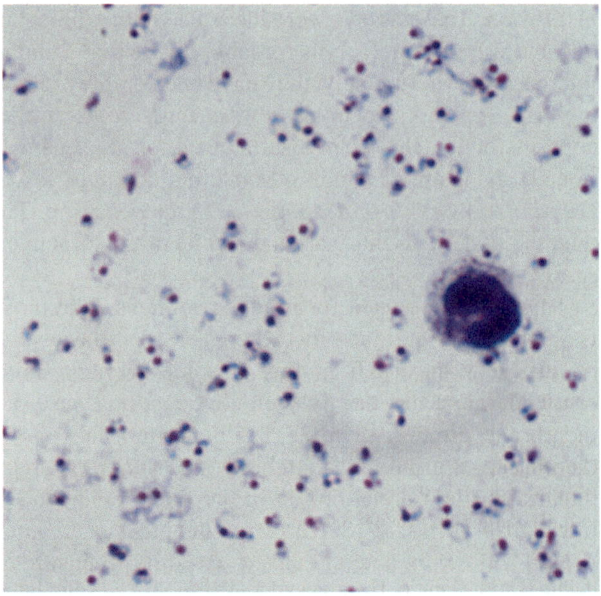

Fig. 4.14 Thick peripheral blood smear stained with Giemsa, showing numerous *Plasmodium falciparum* ring forms (1000× magnification)

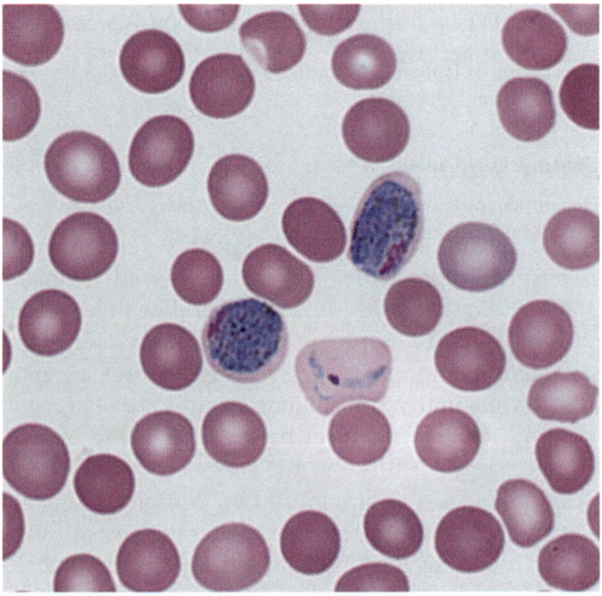

Fig. 4.15 Thin peripheral blood smear showing stained with Giemsa, showing two gametocytes and an amoeboid trophozoite of *Plasmodium vivax* (1000× magnification)

visualized. This number is then multiplied by 100 to convert to a percent parasitemia. Percent parasitemia should always be reported for *Plasmodium* infections and may be reported for *Babesia* though it may not be as meaningful since *Babesia* can have many extracellular forms that will not be appreciated by the parasitemia calculation.

Nucleic Acid Amplification Tests (NAAT)

NAAT testing in parasitology is primarily limited to real-time single target PCR, multiplex "syndromic" PCR, and sequencing-based methods for specification/typing. While the chemistries of these nucleic acid amplification methods differ greatly, the general assay steps are similar: specimen processing to release nucleic acids; amplification of target or probe nucleic acid, or of signal; and detection. However, the performance of molecular assays can vary significantly due to several factors including nucleic acid extraction methods, primer and probe design, amplification and detection technologies, instrumentation, and technical expertise.

The most recognized NAAT is polymerase chain reaction (PCR). The basis of this method is primer-directed amplification of DNA. A pair of primers (short oligonucleotides, usually 20–30 bases in length) is designed so that one primer binds to a complementary section of one strand of a double-stranded DNA molecule, while the other binds to the other DNA strand at a distance of typically 100–300 nucleotides apart (real-time PCR assays are often designed to amplify shorter regions of nucleic acid). In this way, the primers flank a short portion of the pathogen genome and define the region to be amplified. Each primer serves as the point at which new DNA synthesis begins, using a DNA polymerase enzyme. The PCR mixture also contains deoxyribonucleoside triphosphates (dNTPs), the "building blocks" for DNA, and other components, so that a new strand of DNA can be synthesized. PCR requires a series of temperature cycles (thermal cycling), each consisting of a high (~95 °C) temperature and one or two lower temperatures (~50–72 °C). Each temperature cycle completes one DNA replication cycle, in which double-stranded DNA target is melted apart, primers anneal to complementary sequences of each DNA strand, and new DNA strands are synthesized (i.e., primers are extended). The end result is the doubling of the number of DNA copies. PCR results in exponential amplification of target; each round of thermal cycling potentially doubles the amount of the target (amplicon) nucleic acid, so that after 30 cycles as many as a billion copies of newly amplified DNA may exist from a single starting piece of viral genome. The amplified DNA is then readily detected using a labeled DNA probe or fluorescent dye, or by other methods.

These tests are primarily used with tissue, blood, aspirates/fluids, urine/genital swabs, and stool. Organisms that are conventionally detected or identified to a genus and/or species level include: *Plasmodium, Babesia, Leishmania, Toxoplasma,* free-living amoeba (e.g., *Balamuthia, Acanthamoeba, Naegleria*), *Trichomonas,* and intestinal protozoa (e.g., *Cryptosporidium, Entamoeba, Cyclospora,* and *Giardia*). In the case of the gastrointestinal protozoa, these organisms have more recently been detected in multiplex syndromic PCR direct from stool. Tests for these organisms are typically simple to perform and provide a rapid turnaround time. In some cases, the protozoa are tested as a mini-panel of just protozoa, and in other assays these targets are detected simultaneously with gastrointestinal viral and bacterial pathogens. These assays can be formulated in traditional PCR platform format (manual extraction, amplification) or packaged into a cartridge, pouch, or other integrated consumable that allows automated processing from sample input to result output. These assays have been shown to be more sensitive than antigen detection and conventional microscopy, albeit at significantly higher cost.

Case 1

Marc Roger Couturier

A 23-year-old man from California presented to the emergency department in Salt Lake City, Utah with a 3 day history of severe headaches, high fever, and myalgias. Additional symptoms were recalled as mild, including nausea, vomiting, and several bouts of diarrhea. The patient had traveled from Nairobi, Kenya to Salt Lake City, Utah via a connecting flight in Paris. Rather than continue his final flight connection to Tahoe City, California (his intended destination), he deplaned in Utah and sought emergency care out of concerns that he had COVID-19 (SARS-CoV-2). The man had tested negative for COVID-19 by PCR several days before his flight as well as within 24 h of departing Nairobi. He had received full vaccination and booster for COVID-19, typhoid, and yellow fever.

Upon questioning by the physician, the patient revealed that he had been working in the Peace Corps in Burundi for 18 months and had spent the week prior to returning to the United States in Nairobi, Kenya visiting his new girlfriend. The patient admitted to having unprotected sex during that previous week and having hiked extensively in the outdoors, where he experienced multiple insect bites including from midges, flies, and mosquitos. He recalled similar activities and outdoor exposures in Burundi as well. In both Burundi and Kenya, he ate locally prepared foods and drank tap water and ground water filtered through a commercially available ceramic purifier. He denied eating bush meat or game meat.

The patient had a temperature documented at 102.6 °F though his temperature was measured multiple times after admission with mostly 99.9–101 °F. Initial labs were drawn including an HIV antigen-antibody combination test and HIV NAAT, CBC with differential (Table 4.1), blood chemistries (Table 4.1), a rapid malaria test, and blood smears with Giemsa stain to detect blood parasites. The laboratory noted that the blood showed significant hemolysis on initial analysis and the CBC revealed numerous intraerythrocytic inclusions on the automated analyzer. A pathologist reviewed the Wright-stained peripheral blood smear and noted *Plasmodium* parasites in "very many" of the red blood cells (RBCs). The rapid malaria antigen test was positive for the HRP antigen which is specific for *Plasmodium falciparum*. The blood smears were subsequently examined in the microbiology laboratory and thick (Fig. 4.16) and thin (Figs. 4.17 and 4.18)

Table 4.1 Laboratory results at admission to the emergency department

Analyte	Value	Reference range
Platelet	**57 k/µL**[a] ↓	159–439 k/µL
Hemoglobin	**6.1 g/dL**[a] ↓	13.0–17.0 g/dL
Hematocrit	**17.3%**[a] ↓	40.1–52.5%
RBC	**1.98 M/µL**[a] ↓	4.70–6.14 M/µL
WBC	5.22 k/mL	4.30–11.30 k/mL
Differential		
Neutrophil %	62.60%	39.4–72.5%
Lymphocyte %	19.70%	17.6–49.6%
Eosinophil %	1.40%	0.4–6.7%
Monocyte %	6.60%	4.1–12.4%
Basophil %	0.70%	0.3–1.4%
Lactate dehydrogenase	**3122 U/L**[a] ↑	100–253 U/L
Sodium	128 mmol/L	136–144 mmol/L
Potassium	3.7 mmol/L	3.5–5.0 mmol/L
Calcium	**0.99 mmol/L**[a] ↓	1.11–1.30 mmol/L
Glucose	**122 mg/dL**[a] ↑	65–110 md/dL
Lactate	1.20 mmol/L	0.70–2.10 mmol/L
Protein, Total, S/P	**4.7 g/dL**[a] ↓	6.5–8.4 g/dL
Albumin	**2.0 g/dL**[a] ↓	3.5–5.0 g/dL
Bilirubin, Total	**3.6 mg/dL**[a] ↑	0.2–1.4 mg/dL
AST	**269 U/L**[a] ↑	16–40 U/L
ALT	5–60 U/L	43 U/L
Alkaline phosphatase	60 U/L	38–126 U/L

[a] Indicated abnormal values

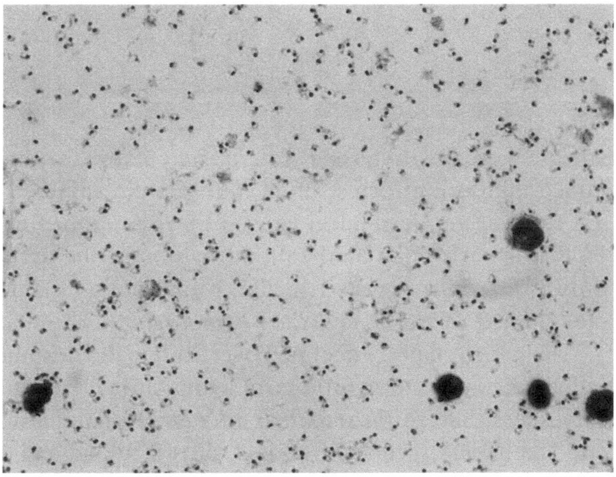

Fig. 4.16 Thick blood smear stained with Giemsa, showing numerous ring form trophozoites of *P. falciparum*. Image captured at 1000× magnification

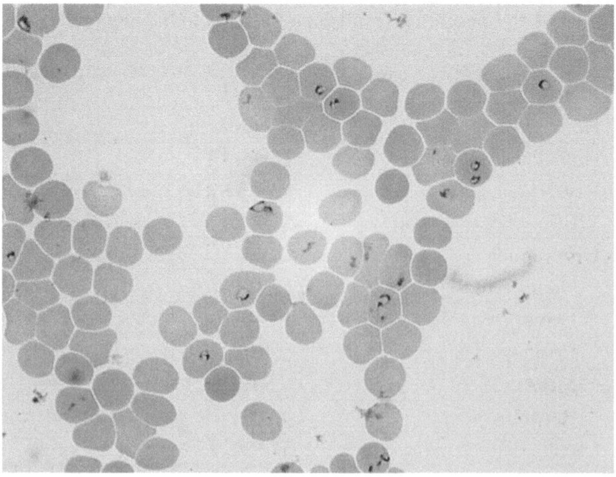

Fig. 4.17 Thin blood smear stained with Giemsa, showing multiple infected red blood cells (black arrow), and "headphone" forms (red arrow) of *P. falciparum*. Image captured at 1000× magnification

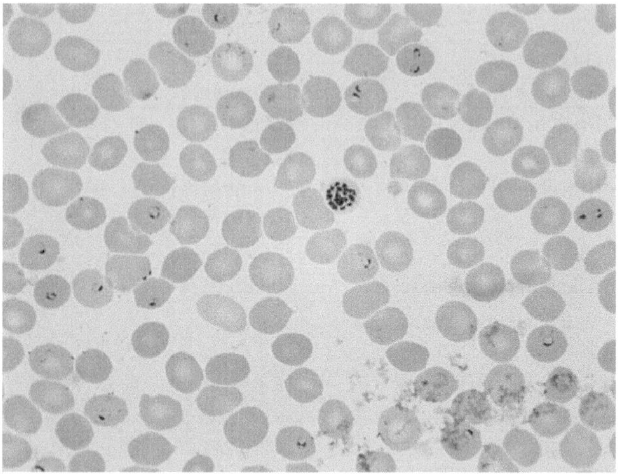

Fig. 4.18 Thin blood smear stained with Giemsa, showing multiple infected red blood cells including a "headphone" form (black arrow), and an appliqué form (red arrow) of *P. falciparum*. A prominent schizont (green arrow) is visible with approximately 14 merozoites and pigment present. Image captured at 1000× magnification

smears were positive for *P. falciparum* at a parasitemia of 20%. The smear contained many ring forms, often multiple per infected RBC and schizonts, which are rarely seen in malaria caused by *P. falciparum* except in severe disease. A PCR was performed on the blood specimen due to the presence of schizonts and only *P. falciparum* was detected, ruling out a mixed infection with another species (i.e., *P. ovale, P. vivax, or P. malariae*).

The patient rapidly declined over 24 h in the hospital and began to show signs of kidney damage and neurological symptoms, including confusion and loss of consciousness. Severe *Plasmodium falciparum* malaria was established as the diagnosis. The patient received multiple units of blood and was treated with intravenous artesunate for 24 h followed by 3 days of artemisinin-based combination therapy consisting of artemether and lumefantrine. The patient improved rapidly after parenteral artesunate and subsequent parasitemias were calculated at 6% after 24 h, 2%

after 48 h, 0.1% after 72 h and 0% after 96 h. The patient remained hospitalized for an additional week to monitor for improvement of kidney function and was eventually discharged without further complications or sequelae.

Discussion

Malaria is caused by the apicomplexan parasite *Plasmodium.* There are four species of human malaria: *P. falciparum*, *P. vivax*, *P. ovale*, and *P. malariae*. The latter is typically characterized by mildly symptomatic illness, with *P. vivax* and *P. ovale* both causing more severe, and sometimes complicated illness. *P. falciparum* is most commonly associated with high parasitemias and severe complications from infections such as major organ failure and cerebral manifestations, especially in children. This species also has the broadest distribution throughout many parts of the world and accounts for the most infections. Though other species of *Plasmodium* may also cause zoonotic/accidental infections in humans (*e.g.*, *P. knowlesi*), they cause malaria in non-human primates and are rarely encountered in human specimens.

The life cycle of human malaria species is largely conserved (Fig. 4.19), with sporozoites entering the human bloodstream from the bite of an infected mosquito and traveling to the liver where they infect hepatocytes. In the liver, a schizont forms and eventually ruptures, releasing merozoites into the bloodstream. The merozoites then infect RBCs and become ring form trophozoites which can either mature into schizonts and release additional merozoites to infect new RBCs or differentiate into gametocytes for sexual reproduction and eventual return to a mosquito via a bloodmeal. The prototypical fever cycles described in malaria are a result of the erythrocytic replication and reinfection of RBCs. One variation in the human phase of *Plasmodium*'s life cycle occurs with both *P. vivax* and *P. ovale*. In these species, a dormant stage known as latent hypnozoites can reside in the liver for weeks, months, and even years. This stage can spontaneously release into the bloodstream resulting in a relapse after successful treatment of the blood phase.

The gold standard method for detection, identification, and enumeration of *Plasmodium* is peripheral blood examination by microscopy. Blood can be stained with Wright, Wright-Giemsa, or Giemsa stains. Giemsa stains are preferred for parasite identification as the staining features are more pronounced for the parasite. Wright and Wright-Giemsa stains can also be interpreted; however, some staining features may not be readily visible and/or sometimes difficult to interpret. A complete blood smear exam includes a minimum of two thin blood smears and two thick blood films. The thick blood films contain significantly more blood volume than the thin films and are lysed, releasing the organisms from the infected RBCs. This film is more sensitive for detecting of parasites though this comes at the expense of losing morphologic features. The thin smears are not lysed and allow for intracellular parasites to be visualized with all morphological features within the parasitized RBCs. The percentage of infected RBCs can also be calculated with this smear.

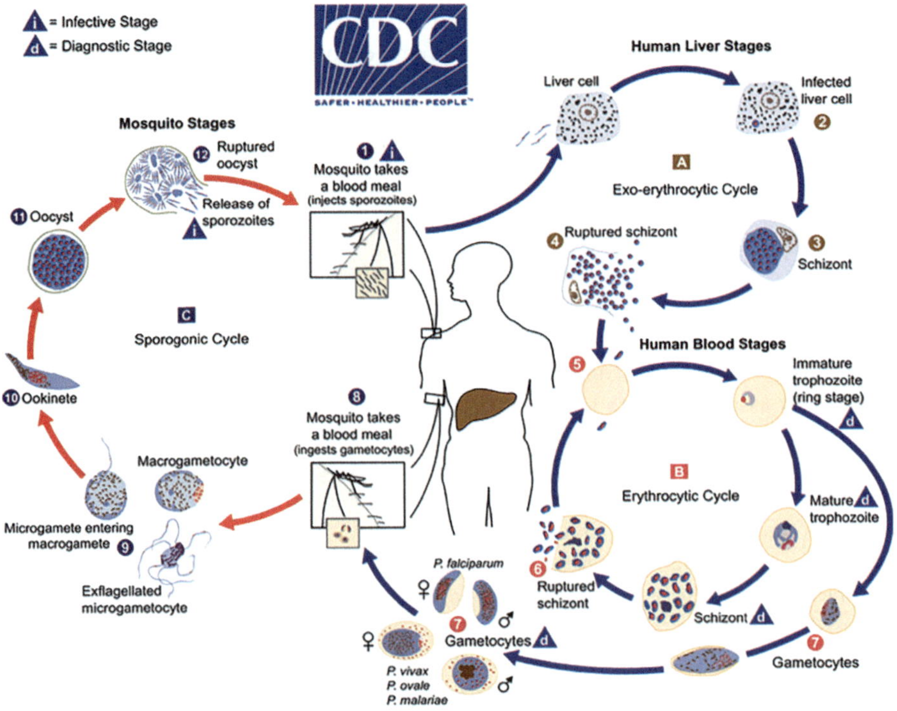

Fig. 4.19 Life cycle of *Plasmodium* species in human and insect hosts. (Image credit to the CDC's DPDx. https://www.cdc.gov/dpdx/malaria/index.html)

P. falciparum is characterized in thin blood films by ring forms infecting all stages of RBCs. RBCs may be infected by multiple ring forms and these rings can display features such as "headphones" or appliqué forms (Fig. 4.17). Headphones refer to rings which display two chromatin dots and the appliqué form designates an elongated/flattened ring on the RBC periphery, sometimes appearing to reside outside of the parasitized RBC. These features are collectively indicative of *P. falciparum*; however, other species can display one or more of these features, so careful examination and consideration of the entirety of features is crucial for establishing an accurate diagnosis. Gametocytes are the only other stage commonly found in peripheral blood, and the morphology is commonly referred to as "banana-shaped." The presence of banana-shaped gametocytes is diagnostic for *P.* falciparum. In rare cases of severe malaria, with hyperparasitemia (generally >10%), other morphologies such as mature trophozoites and schizonts may be seen. In this patient, though no gametocytes were present in the blood film, there were numerous schizonts and mature trophozoites present. Due to the unusual nature of these forms in *P. falciparum* infections, nucleic acid amplification tests (NAAT) could be useful to rule out a dual infection with another *Plasmodium* species. If a NAAT is used to rule out a mixed infection, the laboratory should have validated that the assay can detect two species when they are present at disparate concentrations (e.g., competitive inhibition). Malaria NAAT assays are generally only available in reference or public health laboratories. The turnaround time for these assays is typically not rapid enough to serve as a primary diagnostic method and should instead be reserved for specification or discrepancy analysis. NAAT assays also cannot accurately determine a parasitemia. Tracking parisitemia over time is critical to monitor response to therapy and any potential resistance that may exist.

Rapid diagnostic tests (RDTs) for malaria have become mainstay diagnostics modalities for many hospitals around the world. These tests are available as immunochromatographic assays/lateral flow devices (often referred to as "rapid cards"). Blood is applied to a membrane directly or after a diluent step and allowed to migrate through the membrane. Antibodies that bind a *P. falciparum*-specific protein and/or pan-malarial protein are captured in the flow device and aggregate to make a visible line. Different assays have different targets and interpretations, so laboratories should follow the package insert carefully to ensure an accurate interpretation of the test is made. In general, these tests are quite sensitive and specific, but not as sensitive as a blood smear at very low parasitemias. As a result, it is recommended that an RDT be followed up with a blood smear regardless of whether the result is positive or negative. For positive RDTs, this is crucial since species determination and parasitemia calculations are determined from the blood smear. For negative RDTs, a blood smear allows the lab to rule out a low-level infection or other blood-borne parasites. In resource limited settings, RDTs may be the only modality available for detecting malaria and may be adequate for guiding therapy in endemic areas. RDTs have inherent limitations when testing patients living in *Plasmodium*-endemic areas. Clinical false-negative results may occur in patients who develop immune tolerance to *Plasmodium* and may have low-level parasitemias despite clinical symptoms. Alternatively, a misdiagnosis may be seen in patients with immune tolerance who have parasitemias that are detectable by RDT and who are symptomatic; however, the symptoms are actually attributed to another endemic infection (e.g., typhoid fever, dengue virus, chikungunya virus).

Both RDTs and blood smears should not be considered as rule out tests for malaria when applied to a single blood specimen. If a patient has a high clinical suspicion for malaria and these tests are negative, additional testing on 2–3 subsequent blood specimens every 12–24 h for a total of three blood collections is recommended. Blood collected during or immediately after a fever spike can increase the yield of these tests.

Treatment of severe malaria is critical and can be lifesaving. Expert recommendations include intravenous or intramuscular artesunate for the first 24 h. Once the patient can tolerate oral antimalarial therapy, a combination of antimalarial drugs should be administered for three additional days. The combination should consist of an artemisinin derivative combined with a longer-acting antimalarial with a different mechanism of action. Exchange blood transfusion is no longer indicated for severe malaria infections.

Summary Points

- *Plasmodium falciparum* is capable of causing severe malaria which can result in high parasitemias and death if left untreated.
- Peripheral blood smear exam using thick and thin films is the gold standard laboratory method for diagnosis of malaria.
- Rapid diagnostic tests for malaria are commonly used in many countries and can provide rapid results, with high sensitivity and specificity.
- Positive rapid diagnostic tests should be confirmed with blood film exam in order to definitively identify the infecting species of *Plasmodium* and to establish a parasitemia. Negative RDTs should also be confirmed by blood film since the blood film is more sensitive.

Selected References

Kamaliddin C, Sutherland CJ, Houze S, Cottrell G, Briand V, Castaneda Mogollon D, Pillai DR. The role of ultrasensitive molecular methods for detecting malaria-the broader perspective. Clin Infect Dis. 2021;73(6):e1387–90.

Mathison BA, Pritt BS. Update on malaria diagnostics and test utilization. J Clin Microbiol. 2017;55(7):2009–17.

Mouatcho JC, Goldring JPD. Malaria rapid diagnostic tests: challenges and prospects. J Med Microbiol. 2013;62(Pt 10):1491–505.

Treatment of severe malaria. In: Guidelines for the treatment of malaria. 3rd ed. Geneva: World Health Organization; 2015. https://www.ncbi.nlm.nih.gov/books/NBK294445/.

Case 2

Angela Ma and Marc Roger Couturier

A 32-year-old male with a diagnosis of diffuse large B-cell lymphoma (DLBCL) since 2019 complicated by a positive human immunodeficiency virus (HIV) status (absolute CD4 lymphocytes: 54 cells/μL, undetectable HIV viral load) presented to his regular hematology visit with complaints of watery diarrhea for the past 6 days. The patient described passing three to five non-bloody stools of significant volume daily but no fever, nausea, or vomiting. His hematologist noted cachexia and the patient confirmed he had lost approximately 15–20 lb in the past month but assumed it was a side effect of his DLBCL therapy. The patient also remarked no recent travel or hospitalizations. He had previously been noncompliant with his antiretroviral therapy (ART) due to financial circumstances, but was now regularly taking his medication, which consisted of dolutegravir, emtricitabine, and tenofovir alafenamide. The patient was also on trimethoprim/sulfamethoxazole, acyclovir, and amphotericin prophylaxis.

The care team had concerns for *Clostridioides difficile* colitis, so the patient was admitted for additional testing and nutrition/rehydration replacement therapy. A 10-day course of oral fidaxomicin was prescribed as empiric therapy for *C. difficile* infection (CDI). Initial laboratory tests included a complete blood count (CBC), chemistry panel, bacterial stool culture, and *C. difficile* antigen and toxin test with reflex to PCR detection of the toxin B gene. The CBC did not show significant changes from the patient's previous results obtained 2 weeks ago, and parameters were expectedly decreased due to his DLBCL diagnosis. The patient's sodium levels were slightly decreased at 129 mmol/L (reference range: 136–144 mmol/L) with no additional abnormal values. No growth of enteric pathogens was reported from the bacterial stool culture and the *C. difficile* antigen test was negative for both glutamate dehydrogenase and toxin A/B. After receiving the negative *C. difficile* antigen test results and noting the patient's failure to improve, CDI therapy was discontinued and an infectious disease consult was requested. The patient had now developed a mild fever with continued weight loss since admission. During the consultation, the infectious disease team learned that the patient had emigrated from West Africa 14 years ago. This additional information along with the patient's previous noncompliance with ART brought intestinal parasitic infections to the top of the team's differential diagnosis. CT imaging of the abdomen also revealed mild diffuse, smooth, small intestinal wall thickening (Fig. 4.20). A stool trichrome stain, modified acid-fast stain, and wet mount were ordered. Abundant *Cystoisospora belli* oocysts were observed in the modified acid-fast stain (Fig. 4.21) and unstained wet mount preparation screened with UV fluorescence (Fig. 4.22). Additionally, the trichrome-stained sample was positive for nonpathogenic intestinal amoebae including *Entamoeba coli* and *Endolimax nana*. Considering the patient's HIV and DLBCL diagnoses, he was prescribed a 4-week course of 160 mg/800 mg oral trimethoprim/sulfamethoxazole. His current trimethoprim/sulfamethoxazole

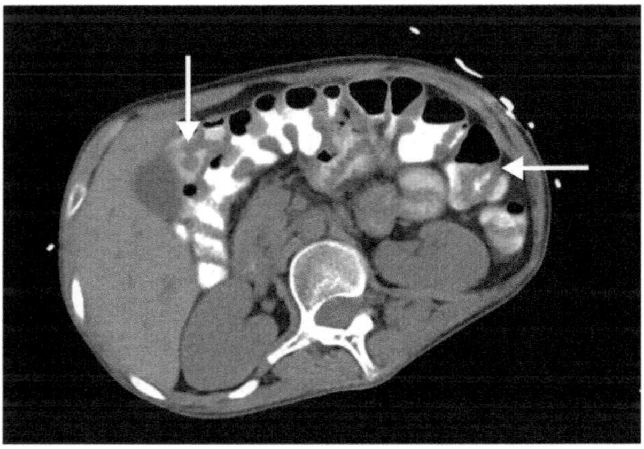

Fig. 4.20 CT image with contrast of patient's abdomen showing mild diffuse, smooth, small intestinal wall thickening (white arrows)

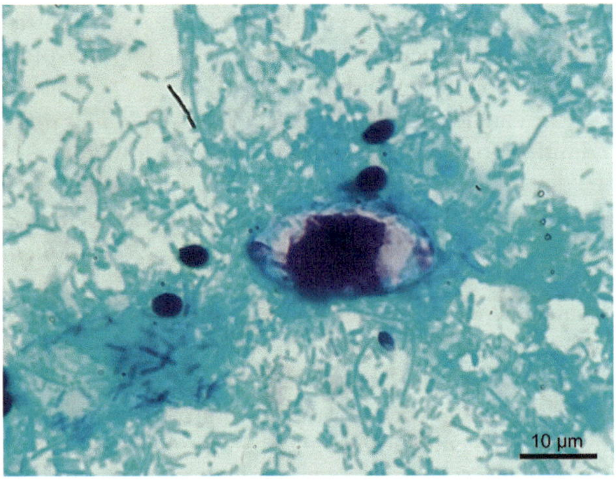

Fig. 4.21 Oocyst of *C. belli* from stool stained with modified acid-fast measuring 25 µm in size, captured at 1000× magnification

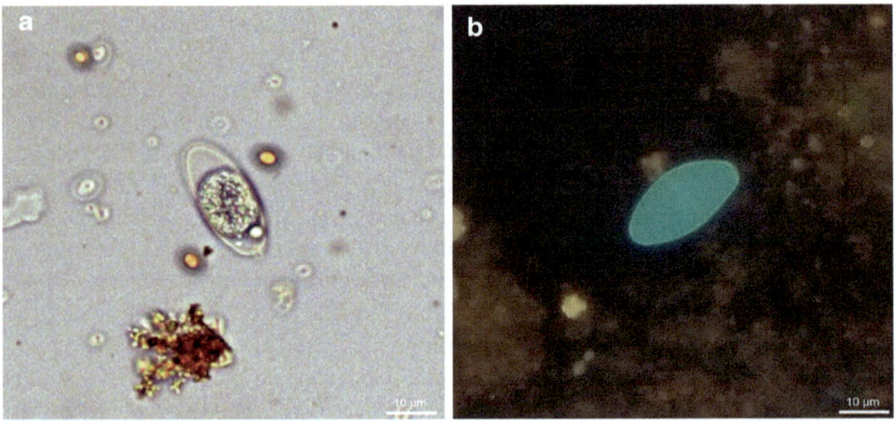

Fig. 4.22 Oocyst of *C. belli* from stool in unstained wet mount under light microscopy (**a**) and ultraviolet fluorescent microscopy (**b**) at 400× magnification

prophylaxis at a dosage of 80 mg/400 mg was believed to be insufficient to treat his cystoisosporiasis. Reactivation of chronic cystoisosporiasis first acquired when the patient resided in West Africa seemed most likely in the case for this patient. The patient's symptoms resolved entirely after completing his course of therapy.

Discussion

Cystoisospora (formerly *Isospora*) *belli* is a coccidian protozoan parasite associated with gastrointestinal infection. Similar to *Cyclospora cayetanensis*, *C. belli* is found worldwide but is more prevalent in tropical and subtropical regions including the

Caribbean, Central and South America, Africa, and Southeast Asia. Patients presenting with cystoisosporiasis from temperate areas most often have relevant travel history to endemic regions or are immunocompromised, and increased prevalence has been specifically documented in AIDS patients. Outbreaks of *C. belli* in the United States are rare but have been reported in institutionalized groups. Humans are currently the only known hosts for *C. belli* with the main route of transmission being fecal-oral.

The life cycle for *C. belli* begins with the fecal excretion of immature, unsporulated oocysts that typically contain a single sporoblast into the environment (Fig. 4.23). Over a duration of 72 h, the sporoblast undergoes a series of divisions and secretion of a cyst wall to produce eight sporozoites per mature oocyst. Ingestion of mature oocysts results in subsequent excystation of the sporozoites in the small intestine. This period of time from excretion of oocysts in feces to ingestion of mature sporulated oocysts is known as the prepatent period, and ranges from 9 to 17 days. Among the coccidian protozoa, *C. belli* sporozoites invade the intestinal epithelial cells the deepest, where they initiate asexual reproduction to produce banana-shaped merozoites that are released into the gut lumen. Merozoites are then able to invade adjacent enterocytes and continue asexual replication or undergo gametogony, where they develop into microgametocytes or macrogametocytes. Fertilization of the macrogametocyte results in conception of a zygote that develops into an immature oocyst, facilitating continuation of the parasite's life cycle.

Infection with *C. belli* is an intracellular process, and the parasite has been detected from various areas of the small intestine, including the duodenum, jejunum, and ileum. Infected patients typically experience a self-limiting watery diarrhea that resolves within 7–10 days. Additional symptoms of cystoisosporiasis include steatorrhea, headache, fever, malaise, and emesis. Infants, children, and immunocompromised individuals may experience more severe symptoms and chronic infection with marked dehydration and weight loss. Furthermore, sporozoites and merozoites are capable of extraintestinal travel to different organ sites to form dormant, monozoic tissue cysts. These thick-walled cysts have been found in the mesenteric lymph nodes, liver, gallbladder, and spleen of patients. In situations where immunity to *C. belli* wanes, monozoic tissue cysts can reactivate, resulting in recurrent infection. The duration of time from when oocysts are first excreted from an infected individual until they are no longer detected is highly variable and referred to as the patent period. Patent periods are heavily influenced by the immune status of the patient, which can last from 30 to 50 days in immunocompetent individuals but often can persist for greater than 6 months in the immunocompromised. First-line therapy for cystoisosporiasis is oral trimethoprim/sulfamethoxazole. For patients with allergies or intolerances to trimethoprim/sulfamethoxazole, ciprofloxacin and pyrimethamine have been used successfully. Patients should also receive fluid, electrolyte or nutrient support as required, particularly in cases of chronic diarrhea and volume depletion.

Due to the similarities in symptoms of cystoisosporiasis with other parasitic and bacterial agents of diarrhea, relevant medical and travel history of the patient is essential. Consecutive stool specimens should be collected over multiple days to increase the probability of microscopic detection of oocysts. When visualized microscopically,

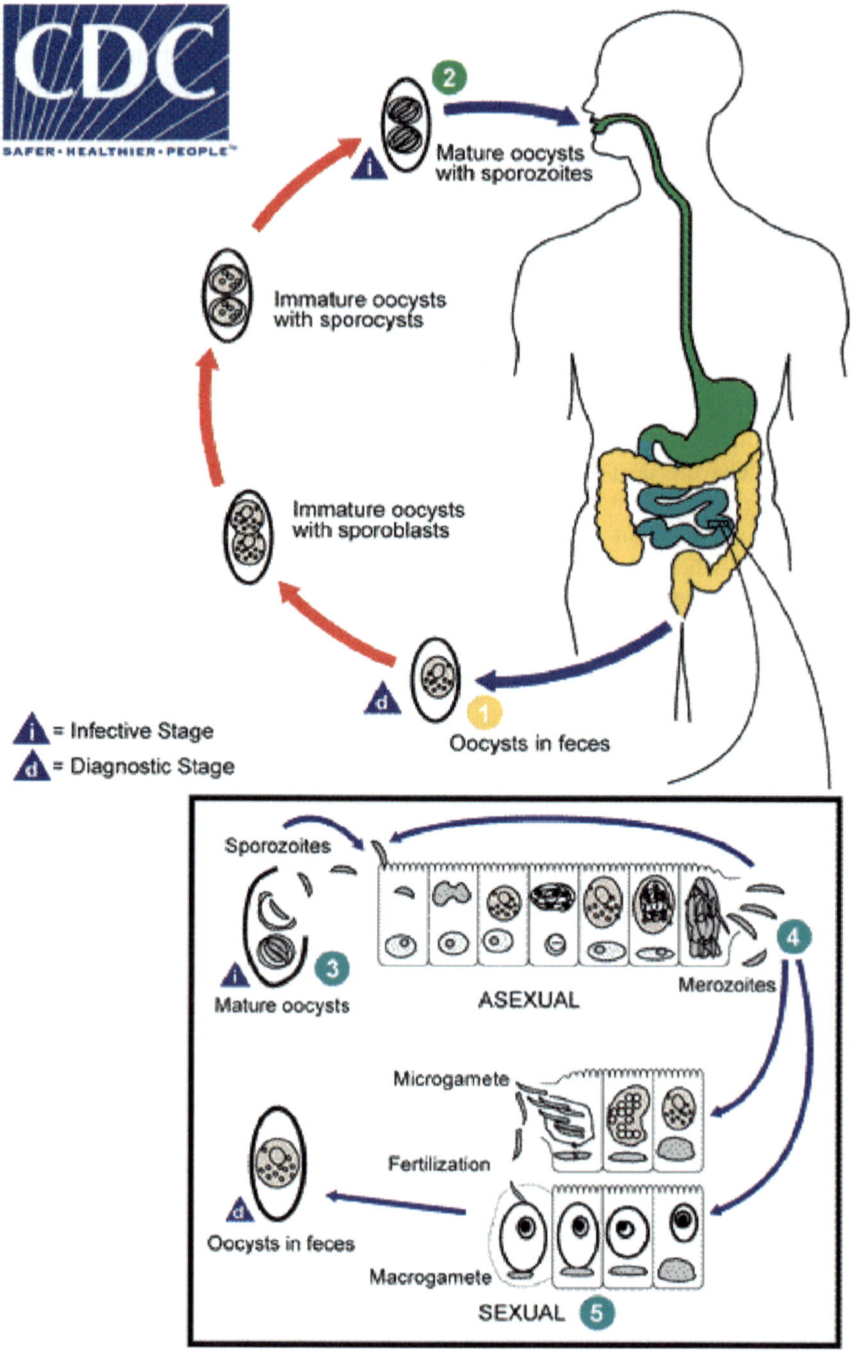

Fig. 4.23 Life cycle of *C. belli* (https://www.cdc.gov/dpdx/cystoisosporiasis/index.html)

C. belli oocysts have an ellipsoidal shape that resembles a football or almond and measure 25–30 μm in size. In stool, oocysts typically contain one or two sporoblasts that appear red against blue-green background material when stained by the cold Kinyoun modified acid-fast or safranin stains. However, *C. belli* oocysts can occasionally fail to uniformly stain and appear as transparent "ghost forms" with the modified acid-fast stain. Similarly, *C. belli* and other coccidian protozoa will appear refractile and transparent if present in stool specimens that have been stained with the trichrome stain commonly used for stool ova and fecal parasite examination. Unstained wet mounts can also be used for diagnosis of cystoisosporiasis, and the oocysts will autofluoresce at excitation wavelengths of 330–365 nm when visualized with ultraviolet fluorescent microscopy (Fig. 4.22). Histopathology of tissue biopsies can demonstrate various developmental stages of *C. belli* within the enterocytes of the small intestine and mucosal changes including shortened villi, crypt hyperplasia, and an increase in inflammatory cells are often observed. In situations of extraintestinal excystation, the presence of sporozoites or merozoites in the form of monozoic tissue cysts can also be observed in tissue sections. In addition to staining with hematoxylin and eosin, *C. belli* is positive with the periodic acid-Schiff stain and resistant to diastase digestion. There are no immunohistochemical stains for detection of *C. belli* in tissue sections. Furthermore, no antigen or serology tests are available for clinical diagnosis of cystoisosporiasis and molecular assays have only been used in research settings.

Summary Points

- *Cystoisospora belli* is the largest of the coccidian protozoa that cause acute to chronic diarrhea in humans, measuring 25–30 μm in size and possessing an ellipsoidal shape.
- There are no animal reservoirs known for *C. belli* and infection is primarily transmitted by fecal-oral routes from contaminated food and water.
- Modified acid-fast or safranin stains and unstained wet-mount preparations of stool are the main methods for detection of *C. belli* infection. There currently are no antigen detection, serology, or molecular assays for diagnosis of cystoisosporiasis.
- Multiple stages of the life cycle of *C. belli* can be visualized by histopathology of small intestine tissue sections when stained with hematoxylin and eosin or periodic acid-Schiff.

Selected References

Centers for Disease Control and Prevention. DPDx—laboratory identification of parasites of public health concern—cystoisosporiasis. https://www.cdc.gov/dpdx/cystoisosporiasis/index.html. Accessed 23 Apr 2022.

Dubey JP, Almeria S. *Cystoisospora belli* infections in humans: the past 100 years. Parasitology. 2019;146(12):1490–527.
Lai KK, Goyne HE, Hernandez-Gonzalo D, et al. *Cystoisospora belli* infection of the gallbladder in immunocompetent patients: a clinicopathologic review of 18 cases. Am J Surg Pathol. 2016;40(8):1070–4.
Lindsay DS, Houk AE, Mitchell SM, Dubey JP. Developmental biology of *Cystoisospora* (Apicomplexa: Sarcocystidae) monozoic tissue cysts. J Parasitol. 2014;100(4):392–8.

Case 3

Cameron A. Brown and James J. Dunn

A 6-year-old male from South Texas presented to the EC with severe frontal headache, fever (38.7 °C), nausea and vomiting, and altered mental status. His symptoms had been getting progressively worse for 3 days prior to presenting at the hospital. He had no significant past medical history and was current on all his vaccinations. The patient's parents reported no sick contacts or recent travel, and family history revealed no significant findings. The parents denied freshwater exposure but did endorse that the child had been at a water park with his older sibling the weekend prior. Given the concerns for meningoencephalitis, the patient was admitted to the pediatric ICU, and started on empiric antibiotic therapy with vancomycin and ceftriaxone.

Examination of the peripheral blood was normal and routine blood cultures were ordered but did not yield any growth. A non-contrast head CT was performed and noted diffuse cerebral edema but was otherwise unimpressive. An external ventricular drain (EVD) was placed for concerns of increased intracranial pressure, and cerebrospinal fluid (CSF) was collected. The opening pressure at EVD placement was noted to be elevated at 50 cm H_2O (reference range: <27 cm H_2O). Increased opening pressure is consistent with the cerebral edema noted on the head CT scan and is observed with many causes of meningitis, including bacterial and tuberculous meningitis.

Given the severity of the presentation and rapid progression of symptoms, the differential diagnosis was primarily focused on bacterial causes of meningoencephalitis, in particular *Streptococcus pneumoniae, Haemophilus influenzae*, and *Neisseria meningitidis,* though *Listeria monocytogenes* remained on the differential despite the family endorsing that the patient had not ingested any unpasteurized dairy products. Several viral etiologies were also considered, including arboviruses such as West Nile virus given the patient presented in the summer. Other viral etiologies considered included enteroviruses and herpes simplex virus 1 and 2, and less likely varicella zoster virus and Epstein-Barr virus. Fungal meningitis, such as *Cryptococcus* or *Histoplasma capsulatum* was less likely given the patient's immunocompetent status.

CSF was sent to microbiology for Gram stain and culture for suspicion of bacterial meningitis. The Gram stain showed no organisms but did reveal many white blood cells. CSF was plated onto routine media, but no organism was isolated. A portion of the CSF was sent to the molecular microbiology lab for a rapid enterovirus reverse transcription-PCR, which was negative.

CSF parameters showed leukocytosis with a white blood cell count of 354 cells/μL (reference range: 0–3 cells/μL) with a neutrophilic predominance (87%) and few red blood cells (24/μL [reference range, 0/μL])), a decrease in glucose (8 mg/dL, [normal range is 50–70% of serum glucose concentration]), and elevated protein (154 mg/dL, reference range: 15–45 mg/dL). The primary care team was called by a hematology technician to report unusual cellular structures found in the Wright-Giemsa-stained CSF fluid that were consistent with trophozoites of an amoeba (Fig. 4.24). The primary care team called the Microbiology lab to see if any motile amoeba could be observed on a wet mount of the CSF specimen, but the specimen had been stored in the 4 °C refrigerator and thus appeared negative.

Given the hematological findings, the infectious disease service was consulted. They recommended a fresh specimen be collected and taken straight to the microbiology lab for another wet mount. With the fresh sample, multiple motile trophozoites could be observed in the specimen, confirming the presence of amoeba in the sample and the diagnosis of primary amoebic meningoencephalitis (PAM) (Fig. 4.25). The patient was given intrathecal amphotericin B and miltefosine was requested from the CDC. The CSF specimen was sent to the CDC for identification, which confirmed that the amoeba was *Naegleria fowleri* by PCR. The patient became comatose and expired the following day.

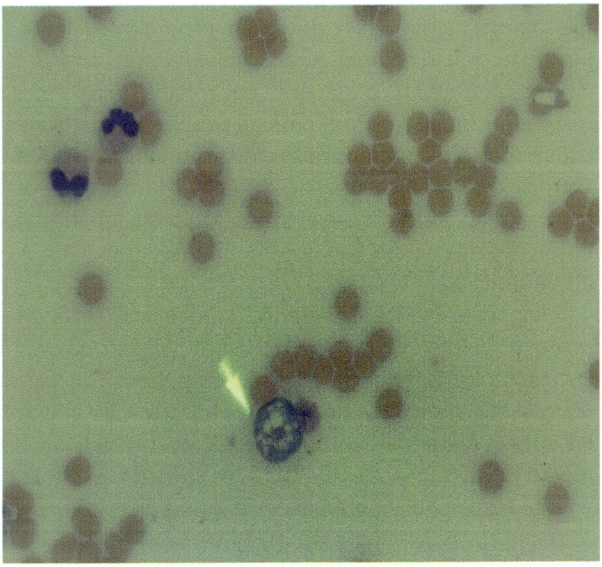

Fig. 4.24 Wright-Giemsa from CSF cell count. The single nucleus and multiple vacuoles can be observed in the trophozoite (green arrow)

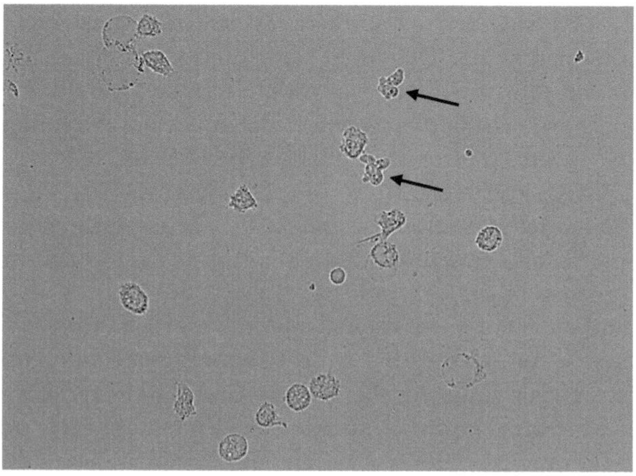

Fig. 4.25 Wet mount of CSF demonstrating multiple trophozoites (arrows)

Discussion

Naegleria fowleri is a free-living amoeba that can be found in warm, freshwater, and moist soil, thriving in hot temperatures up to 45 °C. In addition to naturally occurring bodies of water, *N. folweri* cases have also been linked to under-chlorinated pools and municipal water supply. Like all species of *Naegleria, N. fowleri* predominantly feeds on bacterium though is unique in being the only known human pathogen in the genera. In the United States, the vast majority of infections occur during summer in the southernmost states, Texas and Florida, namely, but sporadic cases have been identified as far north as Minnesota, likely from artificially elevated water temperatures.

CNS infection with *N. fowleri* a occur with contact of the nasal mucosa with water or soil containing the amoebae (Fig. 4.26). From there, *Naegleria* trophozoites can penetrate the cribriform plate to reach the forebrain. Other free-living amoebae, most notably *Acanthamoeba* spp. and *Balamuthia mandrillaris,* which cause granulomatous amoebic encephalitis, can enter the bloodstream through the respiratory tract via nasal passages or through broken skin—CNS invasion can then occur through hematogenous dissemination. Clinical manifestation between the various free-living amoebae differs in terms of severity and timing. *Naegleria* is generally considered to have the shortest incubation period and most severe presentation. Of more than 140 confirmed infections with *N. fowleri* in the United States, all but four were fatal.

Diagnosis for the free-living amoebae is difficult and often missed, while the rarity of the disease and likelihood of *Naegleria* being in the differential means that an antemortem diagnosis is rare. *Naegleria fowleri* may be observed in the CSF using a wet mount to look for trophozoites. Trophozoites will demonstrate characteristic pseudopodal motility, the speed of which can help to distinguish from motile macrophages. Importantly, the CSF should be kept at room temperature or the slide warmed prior to examination in order to visualize motility. *Acanthamoeba* and *Balamuthia* are much less likely to be observed directly in the CSF, but, unlike *Naegleria*, are often detected as cysts in the brain tissue of affected patients. Staining

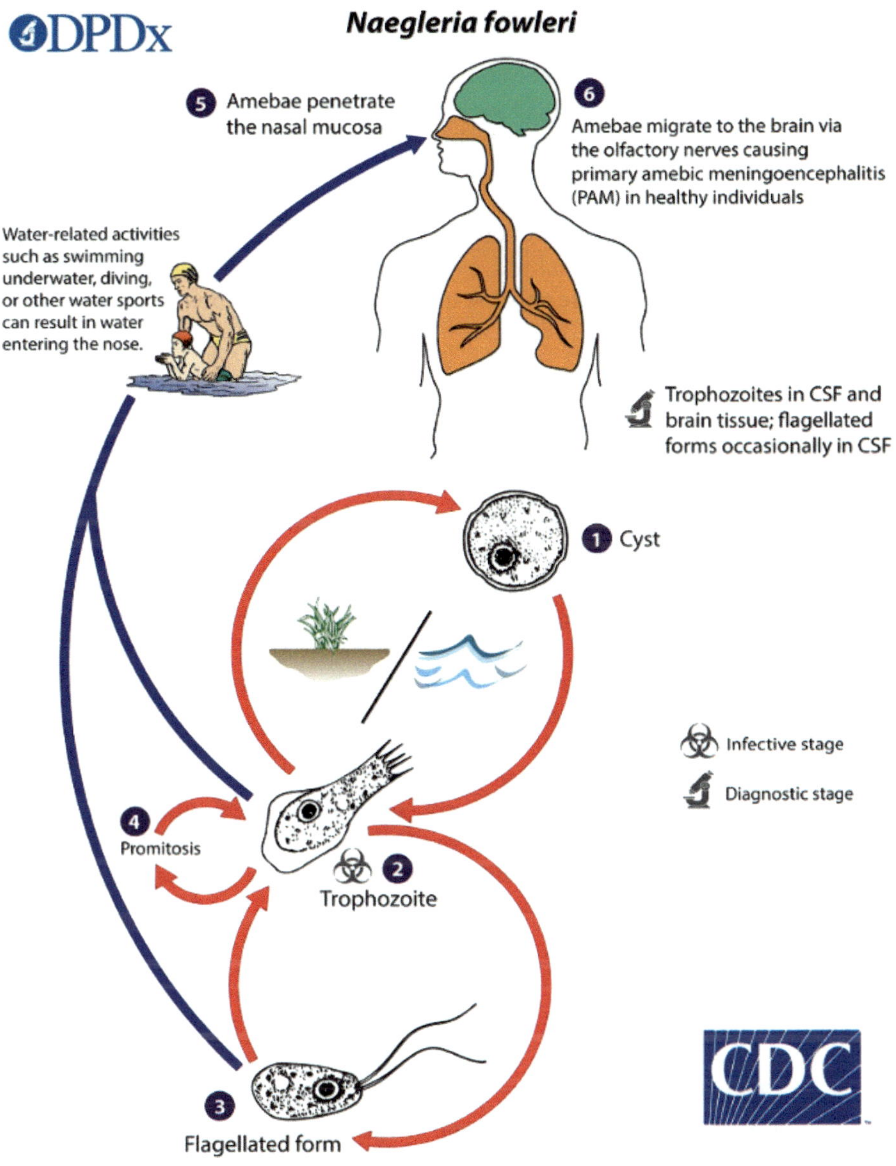

Fig. 4.26 *Naegleria fowleri* life cycle. (Available at https://www.cdc.gov/dpdx/freelivingamebic/index.html)

of the CSF should be done with Giemsa or Wright's stain, which will demonstrate the highly vacuolated nature of the cell alongside with the single nucleus with dense karysome. Gram stain should be avoided due to the possibility of heat fixation destroying the amoebae.

Fluorescent staining of brain tissue, in particular the olfactory bulbs, can be used to confirm the identification of *Acantamoeba, Balamuthia,* or *Naegleria*. A multiplex real-time PCR has been developed at the CDC and other reference laboratories to detect *N. fowleri, B. mandrillaris,* or *Acanthamoeba* spp., which can be used on CSF or tissue samples. For both direct visualization and PCR, it is possible to increase the sensitivity of these assays by attempting to culture *N. fowleri* or *Acanthamoeba* spp. This can be done by adding specimen to a nonnutrient agar plate with a light, uniform suspension of bacteria such as *E. coli* to act as a food source for the amoeba. Tracks in the agar can be observed at microscopically low power as the motile trophozoites consume the bacteria, and trophozoites can be observed at higher power. In culture, cysts, motile trophozoites, and rarely flagellates can all be observed, while only the trophozoites will be observed from a direct patient specimen.

Summary Points

- *Naegleria fowleri* is a rare cause of meningoencephalitis that can resemble acute bacterial meningitis in both clinical presentation and CSF findings. Recent cases due to municipal and underchlorinated water highlight the importance of keeping free-living amoebae in the differential during the summer, even if the patient history does not include freshwater exposure.
- There is no definitive therapy for primary amoebic meningoencephalitis though the few successful treatments included prompt diagnosis and amphothericin B. Miltefosine has been shown to be effective in *Balamuthia* and *Acanthamoeba* infections and may be effective for *N. fowleri.*
- Preliminary diagnosis of primary amoebic meningoencephalitis generally involves observation of amoebae in the CSF, either via wet mount or Giemsa stain. Definitive diagnosis can include specific fluorescent antibody staining of tissue or real-time PCR of tissue or CSF.

Selected References

Capewell LG, Harris AM, Yoder JS, Cope JR, Eddy BA, Roy SL, Visvesvara GS, Fox LM, Beach MJ. Diagnosis, clinical course, and treatment of primary amoebic meningoencephalitis in the United States, 1937–2013. J Pediatr Infect Dis Soc. 2015;4(4):e68–75.

Qvarnstrom Y, Visvesvara GS, Sriram R, da Silva AJ. Multiplex real-time PCR assay for simultaneous detection of Acanthamoeba spp., Balamuthia mandrillaris, and Naegleria fowleri. J Clin Microbiol. 2006;44(10):3589–95. https://doi.org/10.1128/JCM.00875-06.

Case 4

James J. Dunn

A 7-year-old female living in Florida presented to the emergency center with a history of nausea, vomiting, and abdominal pain for 5 days. She had been given intravenous fluids 2 days previously for dehydration but continued to have symptoms. She was recently adopted out of a difficult social situation and had been living in an area with poor sanitation and hygiene. According to her current caregiver "she puts everything in her mouth." She had no history of respiratory symptoms such as cough or wheezing. Results of the physical exam were normal. An abdominal ultrasound was performed and was within normal limits. The patient was admitted to the hospital for fluid resuscitation and supportive care.

The results of a complete blood count obtained at the time of admission showed white blood cell (WBC) count of $9.8 \times 10^3/\mu L$ (normal range, 4.3–11.4 $\times 10^3/\mu L$) with a cell differential of 57% neutrophils (normal range, 28.6–74.5%), 23% lymphocytes (normal range, 15.5–57.8%), 6% monocytes (normal range, 4.2–12.3%), and 11% eosinophils (normal range, 0–4.7%). A blood count performed on the second day of admission showed WBC = $8.6 \times 10^3/\mu L$ with 31% eosinophils. Chemistry testing was unremarkable except for elevated lipase = 745 U/L (normal range, 25–120 U/L) and amylase = 213 U/L (normal range, 30–115 U/L), suggestive of pancreatitis.

A parasitic infection was suspected in this child after finding an unexpected peripheral eosinophilia on two consecutive days. Three stool samples collected on successive days were submitted for laboratory testing. A stool culture from the first specimen was negative for enteric pathogens. Ova and parasite examinations of the first two specimens were negative. The third specimen demonstrated the presence of rhabditiform larvae of *Strongyloides stercoralis* on a direct wet mount (Fig. 4.27). The patient was treated with ivermectin twice daily for 2 days and fully recovered.

Discussion

Strongyloides stercoralis is a soil-transmitted nematode that is prevalent in humid, tropical and subtropical, developing countries but is considered uncommon in developed areas such as Europe and North America. There are areas of the Southeastern United States and Southern Europe where *S. stercoralis* is endemic. The estimated prevalence is between 50 and 100 million infections worldwide. Although rare in developed countries, strongyloidiasis can be a problem in socially and economically marginalized groups and is known as a neglected infection of poverty. A second species known to cause human infection is *S. fuelleborni*. It is less common and found mainly in Africa and Papua New Guinea.

S. stercoralis has two distinct life cycles; the free-living cycle in the environment and the parasitic cycle within the host (Fig. 4.28). Filariform larvae in the soil infect the human host by dermal penetration, usually in the feet or ankles. Larvae enter the bloodstream and are transported to the lungs before ascending the bronchial tree, upon which they are swallowed and enter the gastrointestinal tract. Alternatively, some larvae are believed to migrate to the intestine via penetration of the abdominal viscera or connective tissue. Once in the small intestine, the larvae molt twice and become adult female worms and embed in duodenal mucosa. The female worms are parthenogenetic, a type of asexual reproduction in which female eggs develop into embryos without fertilization. The worms lay embryonated eggs that hatch in the intestine releasing the rhabditiform larvae which can then be passed in stool (Fig. 4.27) or mature into filariform larvae and cause autoinfection. In the latter situation, infective filariform larvae can penetrate the intestinal mucosa or skin of the perianal region and restart the parasitic cycle, potentially leading to hyperinfection syndrome and disseminated disease. In the former situation, rhabditiform larvae passed in stool can become infectious filariform larvae directly or go through a free-living cycle of development in the soil. The rhabditiform larvae of *S. stercoralis* are morphologically similar to those of human hookworm (*Ancylostoma duodenale* and *Necator americanus*), but it's much more likely to detect hookworm eggs in stool rather than larvae unless there is a delay in processing of the sample. *S. stercoralis* and human hookworm larvae can be differentiated only by the morphology of the buccal cavity and genital primordia.

Most infected individuals are asymptomatic or may have intermittent symptoms. Clinical manifestations of *S. stercoralis* infection include irritation at the site of skin penetration, diarrhea, abdominal pain, nausea, cough, wheezing, and shortness of breath. Moderate or marked peripheral blood eosinophilia may be present during the early stages of infection. Loeffler's syndrome, a disease in which eosinophils accumulate in the lung, can be associated with several parasitic infections including *S. stercoralis*. Larva currens associated with infection manifests as pruritic linear

Fig. 4.27 Rhabditiform larva observed in iodine wet mount of stool sample (magnification, 200×)

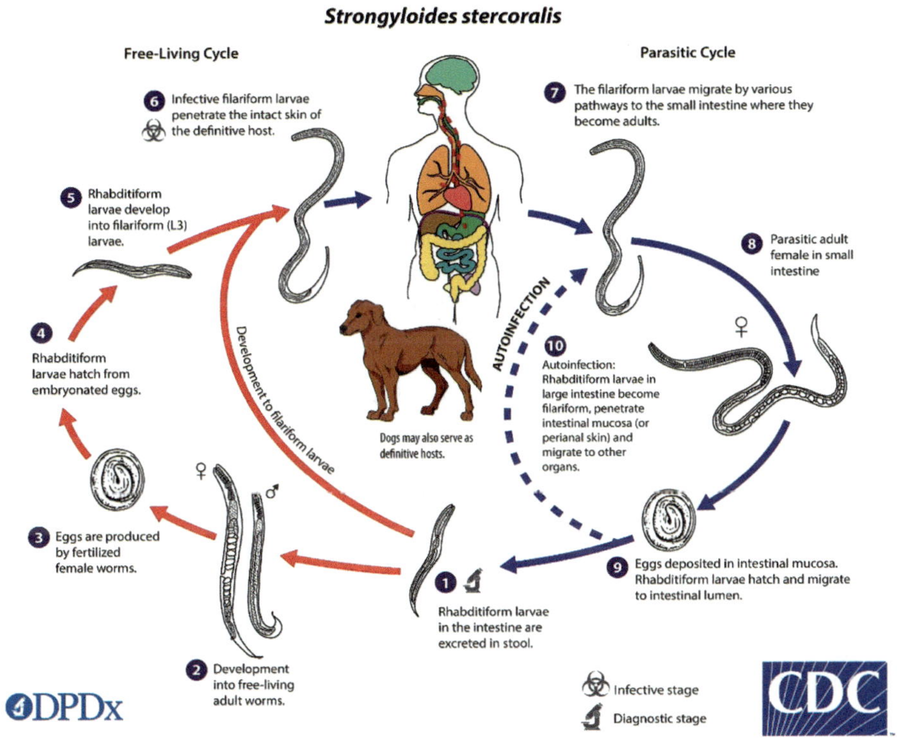

Fig. 4.28 Life cycle of *Strongyloides stercoralis*. (Available at https://www.cdc.gov/dpdx/strongyloidiasis/index.html)

streaks along the lower trunk, thighs, and buttocks from migrating larvae through the subcutaneous tissues. *S. stercoralis* infections often go unrecognized because individuals with chronic infection are frequently asymptomatic or have mild, nonspecific symptoms over the course of many years. Some unusual manifestations of strongyloidiasis include duodenal obstruction, ileus, reactive arthritis, ascites, hepatic lesions, and pancreatitis.

Severe infection can occur in immunocompromised patients or those co-infected with human T-cell lymphotropic virus type 1 (HTLV-1) when hyperinfection and dissemination occur. Hyperinfection is characterized by more pronounced gastrointestinal symptoms and worsening respiratory complaints that can lead to respiratory failure. During hyperinfection, invading filariform larvae transport luminal bacteria to the bloodstream and central nervous system leading to bacteremia and meningitis. Hyperinfection and dissemination can be difficult to treat and mortality rates range from 70% to 85%.

The diagnosis of *S. stercoralis* infection is typically made by detection of rhabditiform or occasionally filariform larvae in the stool, duodenal fluid, and/or biopsy specimens, and possibly sputum or bronchoalveolar lavage (BAL) fluid in disseminated infections. Stool can be examined in direct wet mounts or using a concentration technique. Rhabditiform larvae measure 200–250 µm in length by 14–16 µm in width. They can be distinguished from hookworm larvae by their short buccal chamber (approximately 4 µm in length) and by the presence of a conspicuous genital primordium that is midway between the esophagus and anus. The sensitivity of detection in a single stool sample is only 30–50%, so multiple specimens may be needed to reliably detect the organism. Importantly, use of the formalin-ethyl acetate concentration technique may remove larvae from the sample thereby reducing the sensitivity of detection. PCR assays have also been employed to directly detect the organism in these types of specimens. Antibodies to *S. stercoralis* can be detected in serum using enzyme immunoassay (EIA) or indirect fluorescent antibody (IFA) tests although the presence of antibody does not distinguish between past and current infection and some assays may cross-react in patients with other helminth infections such as filariasis, schistosomiasis, and ascariasis. Culture plates of specimens such as sputum or BAL harboring larvae may show visible tracks created as the organisms carry bacteria over the agar (Fig. 4.13). The larvae can also be identified in permanently stained smears of these respiratory specimens. Standard precautions are used for the processing of stool and respiratory specimens, but raw samples and agar plate cultures should be handled with caution to avoid percutaneous exposure to filariform larvae.

The drug of choice for treatment of strongyloidiasis is ivermectin, which kills the worms in the intestine. It is usually given in two doses 1–14 days apart in uncomplicated cases. In immunocompromised patients with hyperinfection syndrome and disseminated strongyloidiasis, it is necessary to treat with ivermectin daily until stool and/or sputum exams are negative for 2 weeks, in order to prevent recurrence of hyperinfection and kill the worms. Reduction of immunosuppressive therapy is also an important part of treatment when possible.

Summary Points

- *Strongyloides stercoralis* infections are common in tropical and subtropical regions of the world and are also endemic in Southeastern United States and Southern Europe.
- Filariform larvae of *S. stercoralis* in the soil infect human hosts by penetration of intact skin, are transported to body sites like the lungs, swallowed and reach the small bowel where maturation into the adult worm occurs.
- Parthogenetic females embed in the duodenal mucosa and lay embryonated eggs that hatch into rhabditiform larvae which can be passed in stool or mature into filariform larvae that can cause autoinfection.
- Chronic infections with *S. stercoralis* can go unrecognized for years due to lack of specific symptoms or subclinical infection.
- Immunocompromised patients can develop potentially life-threatening hyperinfection syndrome and disseminated infection which is associated with a high mortality rate.

Selected References

Buonfrate D, Requena-Mendez A, Angheben A, Munoz J, Gobbi F, Van Den Ende J, Bisoffi Z. Severe strongyloidiasis: a systematic review of case reports. BMC Infect Dis. 2013;13:78.

Krolewiecki A, Nutman TB. Strongyloidiasis: a neglected tropical disease. Infect Dis Clin North Am. 2019;33:135–51.

Montes M, Sawhney C, Barros N. Strongyloides stercoralis: there but not seen. Curr Opin Infect Dis. 2010;23:500–4.

Nutman TB. Human infection with strongyloides stercoralis and other related strongyloides species. Parasitology. 2017;144:263–73.

Case 5

Angela Ma and Marc Roger Couturier

A 72-year-old female with a recent medical history of colon cancer treatment presented to her gastroenterologist in July with complaints of several weeks of diarrhea. She described the diarrhea as explosive, loose, and voluminous, and estimated she was passing 2–3 stools per day with intense abdominal pain prior. The patient also remarked that the diarrhea typically occurred after eating. She had been experiencing chills and sweats, but no fever. This was the first instance of the patient seeking medical attention regarding her diarrhea, and she had not been taking any antibiotics recently. The patient denied any recent travel outside of Utah since completing her treatment course for colon cancer in February. She further elaborated that she had mostly returned to her old lifestyle pre-colon cancer diagnosis, which consisted of taking care of her grandchildren, gardening, and spending time with her husband. However, the patient did mention that she had started a new diet largely consisting of leafy green salads within the past 2 months after reading an article on the internet.

Upon learning of the patient's diet changes and considering the summer season and duration of symptoms, the physician decided that the patient's presentation did not correlate well with colon cancer relapse or viral gastroenteritis and ultimately were placed lower on the differential. When the patient was asked if she was thoroughly washing the fresh produce used in her salads, she was reluctant to provide an answer. Her physician opted to forgo a colonoscopy and biopsy for the patient's initial visit and ordered bacterial stool culture, modified acid-fast stain, and an intestinal parasites multiplex PCR assay that included *Cryptosporidium hominis* and *C. parvum*, *Cyclospora cayetanensis*, *Dientamoeba fragilis, Entamoeba histolytica,* and *Giardia duodenalis* as targets. The stool culture was negative for bacterial pathogens, but the modified acid-fast stain revealed oocysts averaging 9 μm in size observed in multiple fields (Fig. 4.29) consistent with

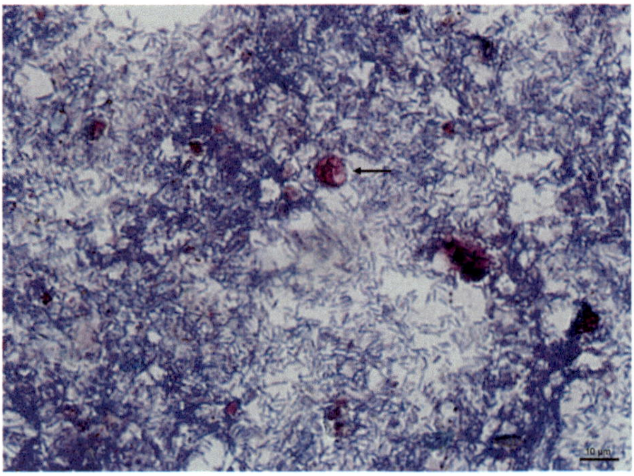

Fig. 4.29 Oocyst of *C. cayetanensis* from stool stained with modified acid-fast denoted by black arrow

C. cayetanensis. Results from the intestinal parasites multiplex PCR that returned 3 days later confirmed the presence of *C. cayetanensis* DNA in the patient's stool. The specimen was negative for all other targets in the intestinal parasites multiplex PCR assay. After receiving these results, the patient was prescribed a 10-day course of oral trimethoprim-sulfamethoxazole taken twice a day and asked to stop her diet of salads. On the patient's follow-up visit, she also admitted to poorly washing her produce and not washing the assorted berries that she would add to the salads. The physician strongly cautioned against not washing fresh produce before consumption and offered to refer the patient to a dietician if she wanted to further discuss her diet plan. After finishing her course of antibiotics, the patient's symptoms resolved entirely.

Discussion

Cyclospora cayetanensis is a coccidian protozoan responsible for cyclosporiasis for which humans are the only known hosts; no definitive animal reservoirs have been identified. *C. cayetanesis* is found worldwide, but more commonly in tropical and subtropical areas including Latin America, the Middle East, Southeast Asia, and Africa. Travelers returning from endemic regions can present with cyclosporiasis following exposure to contaminated food, water, or soil. Infection with *C. cayetanensis* in the United States is most frequently reported in the spring and summer months and is mainly associated with contaminated fresh produce such as raspberries, lettuce, and herbs imported from endemic regions.

The life cycle of *C. cayetanensis* begins with passage of the unsporulated oocyst in stool and subsequent environmental contamination (Fig. 4.30). Sporulation occurs over the course of several days to weeks, where the oocyst enters its infective stage. Ingestion of food or water contaminated with *C. cayetanensis* leads to

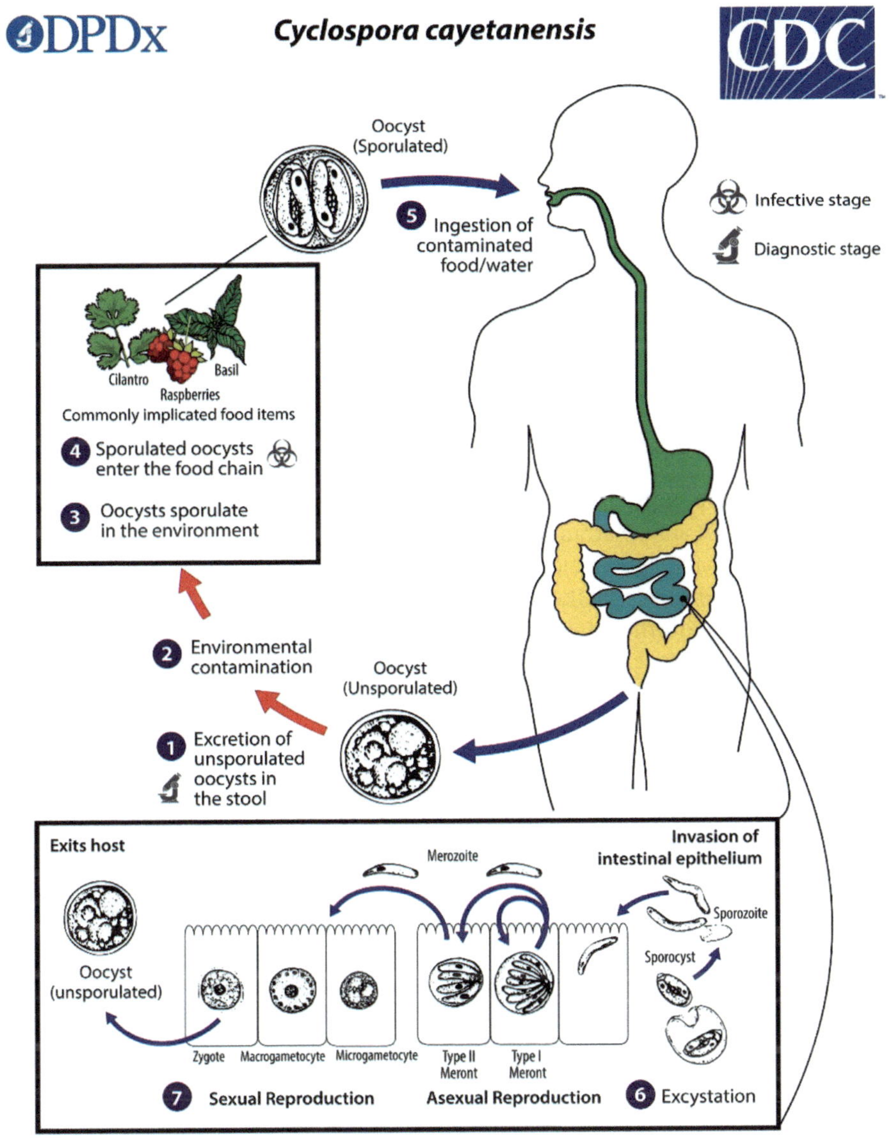

Fig. 4.30 Life cycle of *C. cayetanensis*. (Available at https://www.cdc.gov/dpdx/cyclosporiasis/index.html)

excystation of sporozoites from the oocyst and invasion of enterocytes in the small intestine. Within the enterocytes, sporozoites undergo asexual multiplication into type I and/or type II merozoites, of which the latter can develop into macro- and microgametocytes for sexual reproduction and formation of a zygote. Zygotes then develop into oocysts that are shed in stool to perpetuate the cycle.

Infected patients present with persistent or chronic watery diarrhea and commonly experience weight loss, fatigue, nausea, and vomiting. Those from endemic areas can be asymptomatically infected. Immunocompromised and HIV-positive individuals may suffer more severe complications associated with *C. cayetanensis* infection and are at increased risk of gallbladder invasion and biliary disease. Without treatment, infection can persist for longer than 10 weeks and patients may relapse after a period of symptomatic improvement. Trimethoprim-sulfamethoxazole is the recommended first-line therapy for *C. cayetanensis* infection. For patients with allergies or intolerances to trimethoprim-sulfamethoxazole, ciprofloxacin or nitazoxanide can be used. Prevention of dehydration is generally advised, particularly in pediatric patients.

Relevant patient history, travel, and season of symptom onset are important criteria for consideration of *C. cayetanensis* in the differential. In all situations of suspected intestinal parasitic infection, multiple stool specimens collected in stool fixatives over a period of 5–7 days allow for optimal microscopic diagnosis. Visualization of *C. cayteanensis* oocysts is best achieved with the cold Kinyoun modified acid-fast stain, safranin stain, or unstained wet mount. Oocysts are spherical with a pink to red appearance against a green counterstain and measure approximately 8–10 μm in size with the modified acid-fast stain. However, the parasite does not always stain uniformly and can appear as transparent or colorless "ghost forms" (Fig. 4.31). *C. cayetanensis* oocysts can also be observed in unstained wet mounts with ultraviolet fluorescent microscopy (Fig. 4.32) due to their ability to

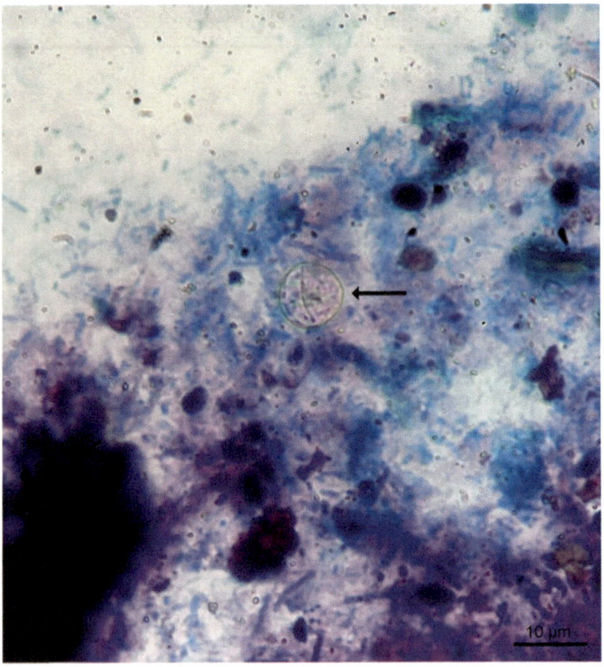

Fig. 4.31 "Ghost form" unstained oocyst of *C. cayetanensis* from stool stained with modified acid-fast denoted by black arrow

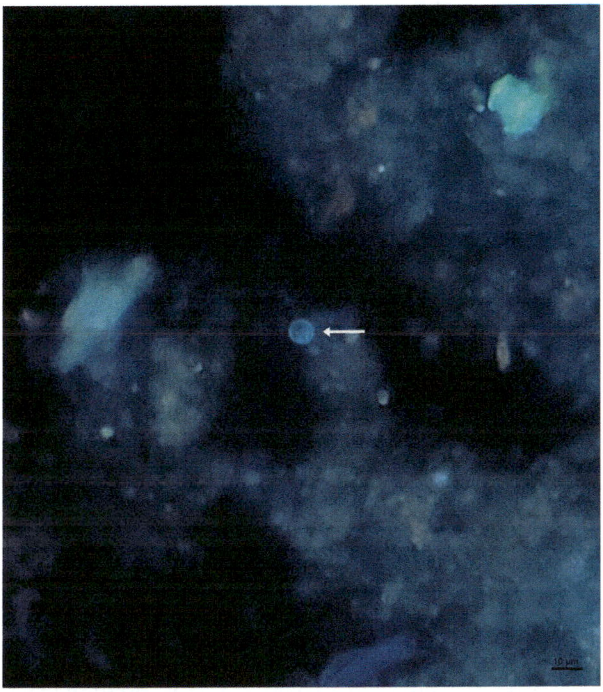

Fig. 4.32 Autofluorescent oocyst of *C. cayetanensis* from stool in unstained wet mount denoted by white arrow

autofluoresce at 330–365 nm. The trichrome stain commonly used in stool ova and parasite examinations may detect *C. cayetanensis* but is not recommended due to its poor ability to stain the parasite. Nucleic acid amplification tests (NAAT) are available to detect *C. cayetanensis* but only one FDA-cleared multiplex gastrointestinal panel includes it as a target. Smaller laboratories may be required to send out specimens to reference laboratories that run the multiplex gastrointestinal panel or have laboratory developed NAAT for *C. cayetanensis*. Several recent outbreaks of *Cyclospora* in the United States have been identified early on due to the broad use of NAAT testing. There are currently no FDA-cleared serology or antigen detection tests available to diagnose cyclosporiasis. Histopathology of small intestine biopsies can also diagnose *C. cayetanensis* infection through identification of the parasite within enterocytes. Hematoxylin and eosin-stained sections can demonstrate the presence of *C. cayetanensis* within vacuoles of the enterocyte above the nucleus. Alterations in the intestinal architecture may also be observed in histopathology, where flattening of the villi and presence of inflammatory cells in the lamina propria are notably increased. Due to the invasive nature of biopsies, histopathology is not routinely performed for diagnosis of cyclosporiasis when noninvasive methods are readily available. However, the parasite can be commonly identified in histopathology following endoscopy and biopsy for suspicion of malignancies in the small intestine. Electron microscopy also has the capacity to identify *C. catenanesis* within enterocytes and can even identify the parasite at various stages of its life cycle, but the method is rarely used clinically.

Summary Points

- *Cyclospora cayetanensis* is a coccidian protozoan that causes intestinal infections and is endemic to tropical and subtropical regions.
- Cyclosporiasis in the United States is largely associated with imported, contaminated fresh produce such as salad greens, berries, and herbs that are consumed from May to August.
- Detection of *C. cayetanensis* can be achieved by light microscopy with modified acid-fast or safranin stains or by fluorescent microscopy in an unstained wet mount to demonstrate autofluorescence.
- Molecular detection of *C. cayetanensis* is a sensitive detection method. Serology or antigen detection tests are not currently available for clinical use.

Selected References

Centers for Disease Control and Prevention. DPDx—laboratory identification of parasites of public health concern—cyclosporiasis. https://www.cdc.gov/dpdx/cyclosporiasis/index.html. Accessed 5 Mar 2022.

Hadjilouka A, Tsaltas D. Cyclospora cayetanensis-major outbreaks from ready to eat fresh fruits and vegetables. Foods. 2020;9(11):1703. https://doi.org/10.3390/foods9111703.

Li J, Cui Z, Qi M, Zhang L. Advances in cyclosporiasis diagnosis and therapeutic intervention. Front Cell Infect Microbiol. 2020;10:43.

Mathison BA, Pritt BS. Cyclosporiasis-updates on clinical presentation, pathology, clinical diagnosis, and treatment. Microorganisms. 2021;9(9):1863.

Case 6

James J. Dunn

Case Presentation

A healthy 11-month-old female was seen by a pediatrician because the mother had noticed the child had some mild perianal pruritus and unusual material in her diaper. These were small, whitish objects approximately 12 mm in length and 3 mm wide. Upon closer inspection by the pediatrician, the objects appeared to demonstrate movement suggestive of a parasite or portion of a parasite that had been excreted by the infant (Video 4.1). The patient displayed no other symptoms and was eating and drinking appropriately. The mother did report that the family had a dog and cat in the household.

The contents of the diaper were placed in a sterile container with saline and sent to the laboratory for further examination. One of the structures was ruptured and the contents stained with iodine. Microscopic examination of a wet-mount preparation of the material revealed numerous large egg packets ranging from 120 to 200 μm in length that each contained 5–15 individual round to oval eggs (Fig. 4.33). The microscopic morphology was unique and characteristic of the intestinal cestode *Dipylidium caninum*, a common parasite found in cats and dogs. The infant had likely incidentally become infected by ingesting dog or cat fleas and, after maturation of the parasite in her small intestine, proglottids, or segments of the worm were being excreted and seen in the diaper. The infant was treated with a single dose of praziquantel and at follow-up 2 weeks later there was no recurrence of proglottids in the stool and the perianal pruritus had resolved.

Discussion

D. caninum is found worldwide and is a common cestode (tapeworm) parasite of dogs and cats and other wild canid and felid species. These animals are the definitive hosts of the parasite while humans are incidental or accidental hosts and only rarely infected. The prevalence of *D. caninum* infection among cats and dogs

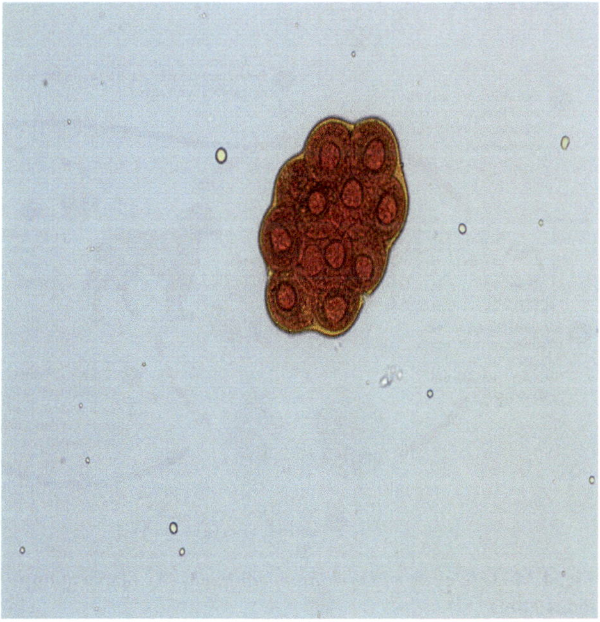

Fig. 4.33 Iodine-stained wet-mount preparation of material expressed from a gravid proglottid. The *D. caninum* egg packets range from 120 to 200 μm in length and house 5–15 individual eggs

ranges from 1% to 60% depending on the geographic location. The life cycle of the tapeworm involves an intermediate host which is usually the dog flea (*Ctenocephalides canis*), cat flea (*Ctenocephalides felis*), or dog louse (*Trichodectes canis*) which has ingested egg packets found in feces in the environment during its larval stage (Fig. 4.34). The parasite eggs are released and mature within the flea or louse and develop into infective cysticercoids. The vertebrate host (animal or human) typically becomes infected by ingesting the adult flea containing the infective cysticercoid which then develops into the adult tapeworm in the small intestine after about 1 month. The adult tapeworms can measure up to 60 cm in length and release individual proglottids which are shed in the stool over time. Excreted, motile proglottids are very common with *D. caninum* infections. They each have characteristic bilateral genital pores through which the egg packets are released.

D. caninum infections in humans occurs most frequently in infants and young children and is often asymptomatic. If symptoms are present they may include anal pruritus, diarrhea, abdominal pain, and appetite loss. Humans serve as a definitive host but *D. caninum* infection is typically not recognized until proglottids are observed in the perianal area, diaper or stool. Laboratory diagnosis is made by visualization of the typical proglottids or characteristic egg packets in stool. Routine ova and parasite testing may not detect the eggs if flotation methods are used since the large egg packets may fail to float. Standard treatment for *D. caninum* infection is a single dose of praziquantel. Infection is best prevented by using flea treatment and deworming of dogs and cats.

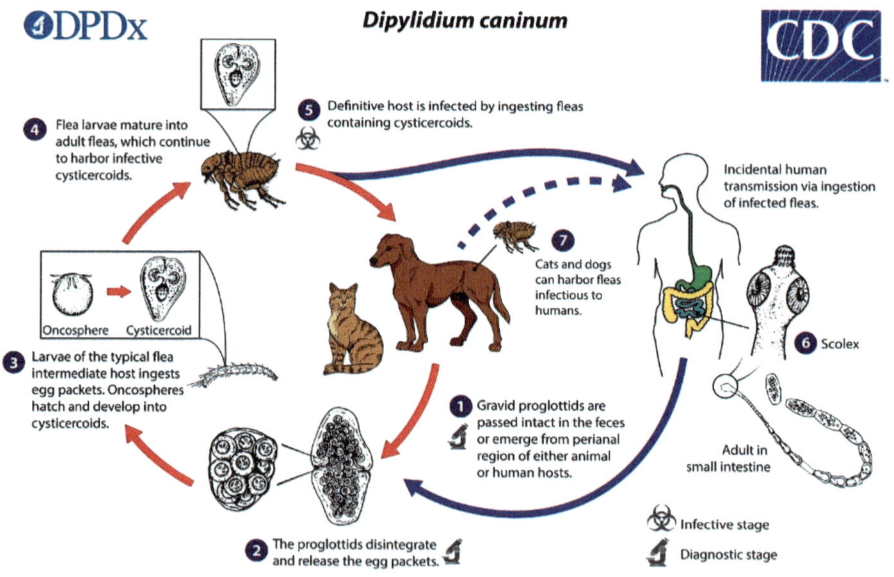

Fig. 4.34 Life cycle of *Dipylidium caninum*. (Available at https://www.cdc.gov/dpdx/dipylidium/index.html)

Summary Points

- Infection with *D. caninum* is a common, worldwide zoonotic disease affecting mostly dogs and cats.
- Humans are not the primary intended host but can be infected with *D. caninum* after ingesting the infected intermediate host—the dog or cat flea.
- Most human infections with *D. caninum* are asymptomatic and occur in infants and young children.
- Recognition and diagnosis of *D. caninum* infection involves observation of the typical proglottids or characteristic egg packets shed in stool.
- Prevention of human infection entails treating dogs and cats with flea repellent and deworming medication.

Selected References

Molina CP, Ogburn J, Adegboyega P. Infection by Dipylidium caninum in an infant. Arch Pathol Lab Med. 2003;127:e157–9.

Portokalidou S, Gkentzi D, Stamouli V, Varvarigou A, Marangos M, Spiliopoulou I, Dimitriou G. Dipylidium caninum infection in children: clinical presentation and therapeutic challenges. Pediatr Infect Dis J. 2019;38:e157–9.

Rousseau J, Castro A, Novo T, Maia C. Dipylidium caninum in the twenty-first century: epidemiologic studies and reported cases in companion animals and humans. Parasit Vectors. 2022;15:131.

Case 7

Santiago Delgado Fernandez and Audrey N. Schuetz

A 67-year-old male from Midwestern United States presented to an outpatient clinic with multiple non-healing cutaneous lesions of approximately 2-month duration. These appeared as painless, crater-like ulcers with central necrosis, and raised, erythematous borders, some with crusted and purulent areas (Fig. 4.35). His wife had similar lesions that had not improved after treatment with antibiotics. The patient was a wildlife photographer and reported recent travel to Costa Rica, where he spent most of his time outdoors, and suffered multiple insect bites. At the clinic, punch biopsies of the lesions were performed, and rapid touch preparations stained with Giemsa were examined (Fig. 4.36). Intracellular ovoid organisms containing a round nucleus and a conspicuous dash-shaped kinetoplast were seen under light microscopy. Tissue sections showed marked dermal lymphohistiocytic inflammation, with *Leishmania* organisms identified within macrophages (Fig. 4.37). Paired with the travel history

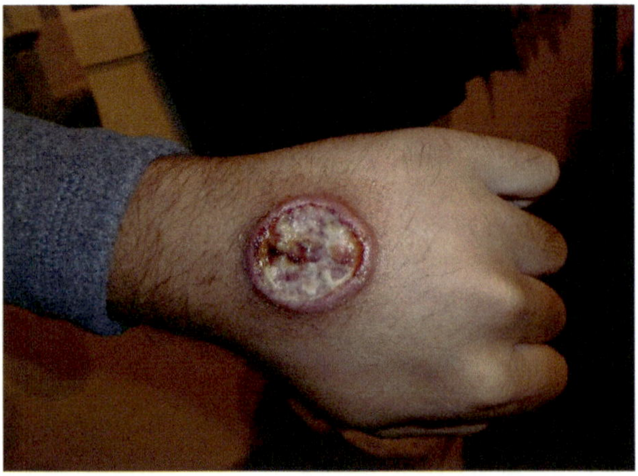

Fig. 4.35 Skin lesion on dorsum of hand showing a crater-like ulcer with central necrosis and a raised, erythematous border, somewhat resembling cheese pizza

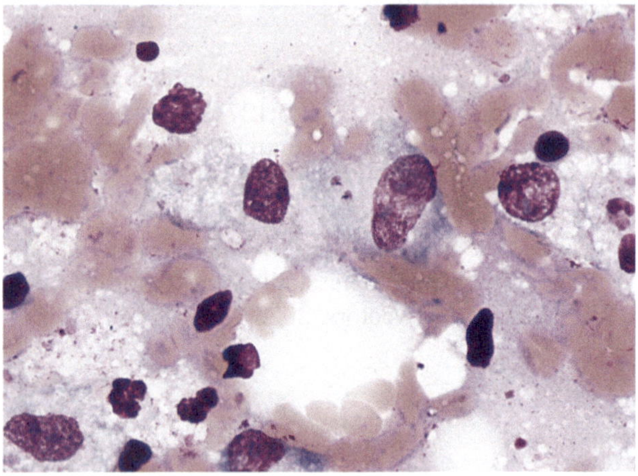

Fig. 4.36 Giemsa-stained touch preparation of skin lesion demonstrating intracellular *Leishmania* organisms with visible bar-shaped kinetoplasts (1000× magnification)

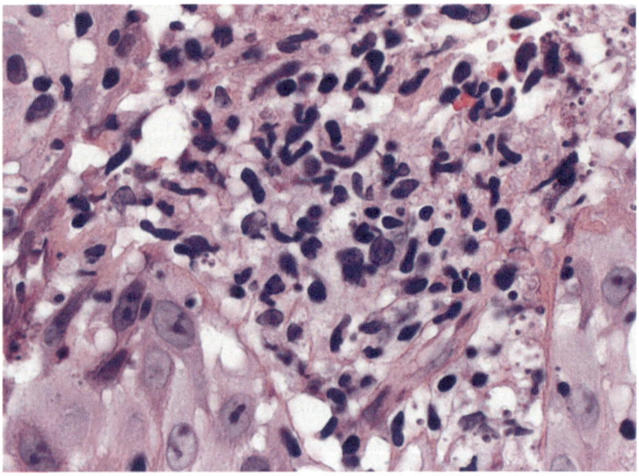

Fig. 4.37 *Leishmania* organisms clustered within vacuoles in this H&E-stained skin biopsy, in a background of lymphohistiocytic and plasmacytic inflammation (1000× magnification)

and the appearance of the skin lesions, a diagnosis of cutaneous leishmaniasis was made. Fresh lesion tissue was sent to the Centers for Disease Control and Prevention (CDC) for culture and species determination by DNA sequencing which are useful in the treatment, prognosis, and epidemiology of the infection. The lesion was positive for *Leishmania panamensis*, a causative agent of cutaneous and mucocutaneous leishmaniasis; based on this result, species-specific treatment was initiated.

Discussion

Leishmaniasis is a vector-borne infection caused by protozoans of the genus *Leishmania*, transmitted by the bite of female sandflies of the genera *Lutzomyia* in the Americas, and *Phlebotomus* in the Old World. Different *Leishmania* species cause a spectrum of disease, and three main clinical presentations have been described: cutaneous, mucocutaneous, and visceral leishmaniasis. Cutaneous leishmaniasis is classically described as a crater-like ulcer with raised borders; this is the mild form of the disease and correlates with an adequate immune response to the parasite. It can be localized (i.e., when a single lesion occurs) or disseminated (i.e., when multiple lesions are present in different parts of the body). The diffuse cutaneous form occurs when there is an anergic immune response which allows for dissemination of the organisms and manifests clinically as hard erythematous nodules and smooth or rough plaques which may become ulcerated. Mucocutaneous leishmaniasis is associated with an exacerbated immune response to the parasite, especially Th1-type and is characterized clinically by lesions of the nasal mucosa that can spread to the oropharyngeal mucosa, with destruction of the tissue and permanent disfiguration. Visceral leishmaniasis, also known as "kala-azar" or "black fever," occurs when the parasite tropism causes it to infect the spleen, liver, and other sites of the reticuloendothelial system, resulting in fever with hepatosplenomegaly, lymphadenopathy, and laboratory abnormalities such as anemia, leukopenia, and thrombocytopenia. The type of disease presentation depends on the infecting *Leishmania* species, which makes species determination a critical step in assessing disease prognosis as well as choice of an appropriate treatment regimen.

The life cycle of the parasite includes the infective form - the motile, flagellated, elongated promastigote found in the sandfly vector - which is injected into the human host through the sandfly's proboscis during a blood meal. In the wound, promastigotes are phagocytized by macrophages and other mononuclear phagocytic cells and transform intracellularly into the nonmotile form, the amastigote. This is the diagnostic stage, morphologically characterized by a nucleus and a dash-like kinetoplast inside a 2–5 μm round structure with no visible flagellum by light microscopy. Amastigotes will multiply intracellularly and invade other phagocytic cells, and sandflies become infected when they ingest these cells during a blood meal. In the vector's gut, the amastigotes will transform once again into promastigotes to complete the cycle.

Diagnosis of the disease relies on epidemiology (e.g., travel history, insect bites), clinical findings, and direct microscopic visualization of the amastigotes in tissue preparations, as illustrated in this case. Touch preparations are often useful in identifying the organisms, as they can be harder to perceive in tissue sections. *Leishmania* amastigotes have been shown to be positive by CD1a immunohistochemical stain. The differential diagnosis includes *Trypanosoma cruzi*, of which the amastigote form is indistinguishable by morphology, but clinical history and location of lesions play a key role. Chagas disease affects mainly cardiac and smooth muscle in the heart and gastrointestinal tract, and skin lesions are not common. *Histoplasma* is a morphologically similar, intracellular round structure that measures 1–5 μm and has a visible nucleus; the lack of a kinetoplast differentiates it from *Leishmania*. The kinetoplast is not always visible, especially on hematoxylin and eosin (H&E) stain, which can make the distinction difficult; in these cases, a Grocott methenamine silver (GMS) stain can be performed which highlights fungi, while *Leishmania* organisms are negative by GMS. Serologic testing is most useful in visceral disease when lesions are not readily accessible for sampling; an IgG host response is also predominantly seen in the visceral type of the infection. Molecular techniques, such as PCR and sequencing analysis, or—historically—cultivation of the organism, are essential for species identification since amastigotes of all species are morphologically identical. Species identification is of paramount importance for treatment choice, risk progression assessment, and epidemiology and control of the disease; it is especially useful when there is a complex travel history to countries that are endemic for multiple *Leishmania* species, as evidenced in the above case. If patients acquired cutaneous leishmaniasis in Central or South America, it is essential to perform species identification to rule out one of the species (e.g., *L. braziliensis* complex) that causes mucocutaneous disease. Complications of mucocutaneous leishmaniasis can be serious, and patients with this form of disease are often treated with systemic therapy rather than topical or oral therapy. Species identification can also be helpful in determining the presence of species which are considered more aggressive than others and are more likely to result in treatment failure.

Summary Points

- Leishmaniasis has three main clinical presentations: cutaneous, mucocutaneous, and visceral.
- The infecting *Leishmania* species and subsequent immune response to the infection play a role in the lesion development and disease progression.
- Diagnosing leishmaniasis entails combining epidemiologic, clinical, and laboratory findings.
- Amastigotes are the diagnostic stage of the parasite and can be seen in touch preparations and biopsy specimens.

Selected References

Aronson N, Herwaldt BL, Libman M, Pearson R, Lopez-Velez R, Weina P, Carvalho EM, Ephros M, Jeronimo S, Magill A. Diagnosis and treatment of leishmaniasis: clinical practice guidelines by the Infectious Diseases Society of America (IDSA) and the American Society of Tropical Medicine and Hygiene (ASTMH). Clin Infect Dis. 2016;63(12):1539–57.

Burza S, Croft, SL, Boelaert M. Leishmaniasis. Lancet. 2018;392(10151):951–70.

Scott P, Novais FO. Cutaneous leishmaniasis: immune responses in protection and pathogenesis. Nat Rev Immunol. 2016;16(9):581–92.

Torres-Guerrero E, Quintanilla-Cedillo MR, Ruiz-Esmenjaud J, Arenas R. Leishmaniasis: a review. F1000Res. 2017;6:750.

Chapter 5
Rare and Notable Cases

Marc Roger Couturier, Audrey N. Schuetz, Santiago Delgado Fernandez, and James J. Dunn

Case 1

James J. Dunn

A 14-year-old male who recently emigrated from Mexico presented to the emergency department with fever and cough that had been progressively been getting worse over the previous month. He was seen as an outpatient after initial onset of symptoms and prescribed antibiotics for a presumed sinusitis infection. One-year prior, the patient had been diagnosed with autoimmune hepatitis and was currently taking oral immunosuppressive therapy while awaiting a liver transplant. His history was significant for possible exposures to unpasteurized milk, mosquito bites and was around some farm animals when residing in rural Mexico. He had no known exposures to anyone with tuberculosis.

On physical exam, he was febrile and the most notable finding was his distended abdomen and a markedly enlarged spleen. Blood tests showed decreased production

M. R. Couturier (✉)
Department of Pathology, University of Utah/ARUP Laboratories, Salt Lake City, UT, USA
e-mail: Marc.Couturier@aruplab.com

A. N. Schuetz
Department of Laboratory Medicine and Pathology, Mayo Clinic, Rochester, MN, USA
e-mail: Schuetz.Audrey@mayo.edu

S. D. Fernandez
Caracas, Venezuela

J. J. Dunn
Department of Pathology, Texas Children's Hospital, Houston, TX, USA

Department of Pathology & Immunology, Baylor College of Medicine, Houston, TX, USA
e-mail: jjdunn@texaschildrens.org

J. J. Dunn et al., *Challenging Cases in Diagnostic Clinical Microbiology*,
https://doi.org/10.1007/978-3-031-65092-5_5

of white blood cells (WBC) and red blood cells (RBC), increased values for inflammatory markers ESR and CRP, abnormal liver function enzymes (AST and ALT), and dysregulation of coagulation factors (PT, INR, PTT, fibrinogen, and D-dimer). Laboratory test values are shown in Table 5.1.

An MRI of the abdomen revealed a massively enlarged spleen with a craniocaudad dimension of 30 cm (normal approximately 10–13 cm). A large lesion was noted toward the posterior superior aspect of the spleen which measured about 4 cm × 3.5 cm (Fig. 5.1). The patient was taken to surgery where aspiration and drainage of the lesion was performed. Approximately 40 mL of purulent fluid was aspirated and delivered to the laboratory for bacterial, fungal, and AFB cultures.

An oxidase-positive, indole-negative Gram-negative bacillus was recovered in the aerobic bacterial culture (Fig. 5.2). The organism was identified as *Burkholderia thailandensis* using MALDI-TOF mass spectrometry. At the time there were scant reports of this organism causing human infection so there was concern that this identification might not be correct. The isolate was tested using two different biochemical identification systems. One automated biochemical system identified the organism as *Pseudomonas aeruginosa* and another manual system identified it as *Burkholderia cepacia* complex. To determine a more accurate identification, segments of the bacterial 16S rRNA gene were sequenced using a pyrosequencing technique. In this process, the nucleic acid of the organism is extracted and purified, subjected to PCR specific for the 16S rRNA gene segments of interest and then those amplification products are analyzed using a pyrosequencing instrument. In this case, the V1, V3, and V6 regions of the 16S rRNA gene were assessed (Fig. 5.3) and based on the data for two of those regions the organism was identified as either *Burkholderia mallei* or *Burkholderia pseudomallei* (Table 5.2). Confirmatory PCR testing performed at the state public health laboratory identified this organism as *B. pseudomallei.*

Table 5.1 Laboratory test results

Parameter	Test value	Reference range
WBC	$3.32 \times 10^3/\mu L$	$4.5–13.5 \times 10^3/\mu L$
WBC cell differential	64% neutrophils	33–76% neutrophils
	14% bands	0–1% bands
	17% lymphocytes	15–55% lymphocytes
	5% monocytes	0–4% monocytes
RBC	$2.73 \times 10^6/\mu L$	$4.5–5.3 \times 10^6/\mu L$
Aspartate transaminase (AST)	56 U/L	10–45 U/L
Alanine transaminase (ALT)	59 U/L	10–40 U/L
Alkaline phosphatase	193 U/L	58–237 U/L
Erythrocyte sedimentation rate (ESR)	67 mm/h	0–20 mm/h
C-reactive protein (CRP)	6.9 mg/dL	<1.0 mg/dL
Prothrombin time (PT)	18.4 s	10.4–15.0 s
International normalized ratio (INR)	1.6	0.8–1.2
Partial thromboplastin time (PTT)	40.4 s	24.9–34.1 s
Fibrinogen	350 mg/dL	220–440 mg/dL
D-dimer	6.06 μg/mL	<0.4 μg/mL

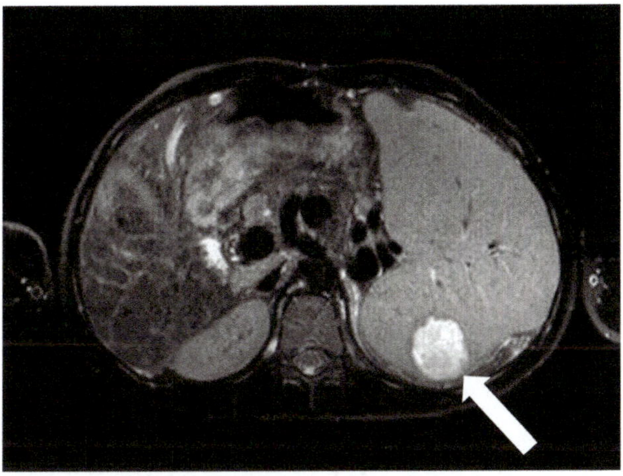

Fig. 5.1 Abdominal MRI showing large splenic lesion toward the posterior superior aspect measuring approximately 4 cm × 3.5 cm (arrow)

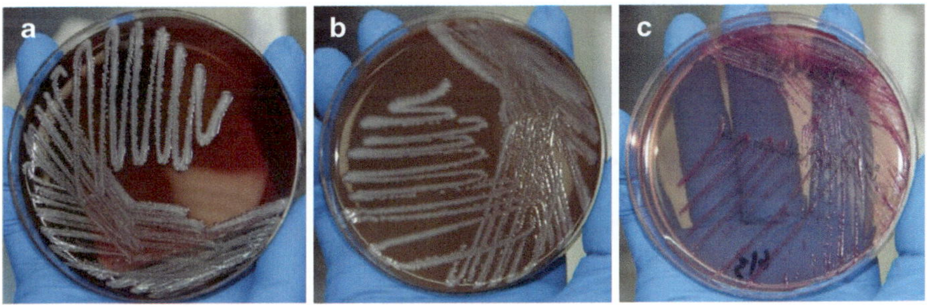

Fig. 5.2 Culture plates showing good growth of the organism within 48 h on sheep blood (**a**), chocolate (**b**) and MacConkey (**c**) agars. The organism was oxidase-positive, indole-negative and could grow at 42 °C

Discussion

Disease caused by *B. pseudomallei* can affect both humans and animals and is called melioidosis. *B. mallei* is the etiologic agent of glanders that primarily affects livestock but can less commonly affect humans. Both *B. mallei* and *B. pseudomallei* are considered overlap select agents which means that they are considered biothreat agents that can affect both humans and animals. In order to identify human infections with these organisms, it is typically up to front-line hospital clinical laboratories (the so-called sentinel laboratories) in the United States to have protocols in place to recognize and rule out other possible etiologies and then refer any suspicious cultures to the public health laboratory for confirmation. Biosafety is critical when *B. pseudomallei* or *B. mallei* is suspected. All cultures should be handled with enhanced biosafety protocols such as biosafety level (BSL)-3 practices.

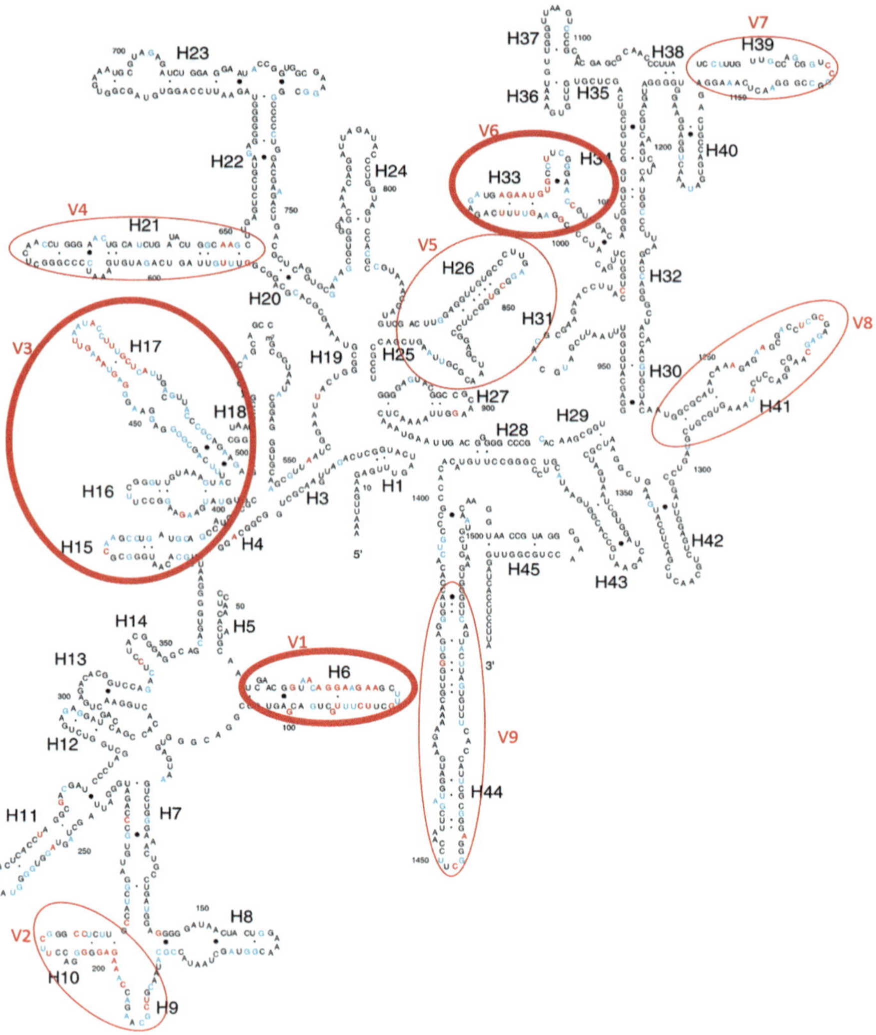

Fig. 5.3 Bacterial 16S rRNA gene. Regions sequenced to determine organism identification included V1, V3, and V6. (Adapted from Case RJ, et al. Appl Environ Microbiol. 2007;73(1):278–88 and Hall JR, et al. Appl Environ Microbiol. 2008;74(15):4910–22)

Melioidosis is endemic in Southeast Asia and northern Australia and sporadic infections have been identified in parts of the Indian subcontinent, Africa, and in Central and South America. Infection with *B. pseudomallei* is typically acquired through direct contact with the environment where the organism persists, such as soil or water. Acquisition can occur by ingestion, percutaneous inoculation or inhalation. The incubation period is usually about 7–10 days. However, in rare cases the

Table 5.2 Nucleic acid sequence database results from pyrosequencing data

16S rRNA region and Organism matches	No. of nucleotide bases (Identity match)
16S rRNA V3 region (44 nucleotides)	
Burkholderia mallei	44 (100%)
Burkholderia pseudomallei	44 (100%)
16S rRNA V6 region (40 nucleotides)	
Burkholderia mallei	40 (100%)
Burkholderia pseudomallei	40 (100%)

organism may remain latent after initial infection and disease may not manifest until decades later. Pneumonia is the most common clinical manifestation but the organism can also enter the bloodstream and cause disseminated infection, skin lesions and ulcers, gastrointestinal tract ulceration and abscesses of the spleen, liver, and kidneys. Initial antimicrobial treatment includes combination therapy with ceftazidime and trimethoprim-sulfamethoxazole for 2–8 weeks. Following treatment of the acute disease, eradication therapy with trimethoprim-sulfamethoxazole alone is indicated for at least 3 months.

In terms of laboratory testing, partial sequencing of the bacterial 16S rRNA gene is an important method used in the clinical microbiology laboratory for those organisms that are difficult to identify accurately by more commonly used methods such as MALDI-TOF or biochemical identification systems. This is particularly true for organisms such as *B. pseudomallei* which are not included in the FDA-cleared identification databases for most commercial systems. The 16S rRNA gene contains alternating regions of sequence conservation and variable regions with less sequence homology which makes it well suited as a target for sequence analysis. The areas of diversity or variable regions are often unique to a particular genus or species. Once sequence is generated the data is compared to a database containing sequences of known organisms to ascertain the identification of the organism in question.

The pyrosequencing method that was employed in this case is one that is useful for short-read-length sequencing for identification of bacteria, fungi, and viruses. It is based on the luminometric detection of pyrophosphate that is generated during DNA synthesis. For bacterial 16S rRNA sequencing, once the variable regions of interest have been subjected to PCR, the amplicons are incubated with a mixture of enzymes and reagents which include DNA polymerase, ATP sulfurylase, luciferase, apyrase, adenosine 5′ phosphosulfate, and luciferin. When deoxynucleoside triphosphates (dNTPs) are added individually and incorporated into the DNA strand synthesized by DNA polymerase a pyrophosphate (PP_i) molecule is released in a quantity equimolar to the amount of incorporated nucleotide. The ATP sulfurylase quantitatively converts PP_i to ATP in the presence of adenosine 5′ phosphosulfate which then drives the luciferase-mediated conversion of luciferin to oxyluciferin generating a light signal captured by the camera in the instrument. The process is repeated by adding the next dNTP.

Summary Points

- Infection with *B. pseudomallei* causes melioidosis in humans and animals after exposure to the environmental reservoir where the organism persists.
- Melioidosis is endemic in Southeast Asia and northern Australia and increasingly recognized to cause sporadic infections in parts of India, Africa, and Central and South America.
- Due to its potential to cause devastating disease in humans and animals, *B. pseudomallei* is considered a potential biothreat agent in the United States.
- Nucleic acid sequencing methods are an important tool in the clinical microbiology laboratory to identify uncommon organisms like *B. pseudomallei* that may not be included in biochemical or MALDI-TOF identification databases.

Selected References

Case RJ, Boucher Y, Dahllof I, Holmstrom C, Doolittle WF, Kjelleberg S. Use of 16S rRNA and rpoB genes as molecular markers for microbial ecology studies. Appl Environ Microbiol. 2007;73:278–88.

Cheng AC, Currie BJ. Melioidosis: epidemiology, pathophysiology, and management. Clin Microbiol Rev. 2005;18:383–416.

Hall JR, Mitchell KR, Jackson-Weaver O, Kooser AS, Cron BR, Crossey LJ, Takacs-Vesbach CD. Molecular characterization of the diversity and distribution of a thermal spring microbial community by using rRNA and metabolic genes. Appl Environ Microbiol. 2008;74:4910–22.

Inglis TJJ, Rolim DB, De Queiroz Sousa A. Melioidosis in the Americas. Am J Trop Med Hyg. 2006;75:947–54.

McLeod C, Morris PS, Bauert PA, Kilburn CJ, Ward LM, Baird RW, Currie BJ. Clinical presentation and medical management of melioidosis in children: a 24-year prospective study in the Northern Territory of Australia and review of the literature. Clin Infect Dis. 2015;60:21–6.

Case 2

Marc Roger Couturier

A 6-year-old female from northern Utah presented to the emergency department at the children's hospital with a rapidly progressing, ascending full-body paralysis. The patient was previously healthy and her parents did not notice any significant symptoms preceding this paralysis. The parents denied any trauma, rash, or elevated temperatures. The patient was minimally communicative but described feeling tired and "lazy" the day before symptoms began though that was attributed to a week of outdoor activities in the prior days. The patient remained afebrile and fatigued, so neurology and infectious disease consults were sought.

The patient and her family had been socially distancing since March 15, 2020 due to the state-mandated COVID-19 response. The family were avid hikers and campers and decided to spend more time in the mountains when the spring snow-pack had melted. The most recent camping trip was a 1-week trip to the western slope of the Manti-La Sal National Forest in central Utah. The family stayed in a pull-behind trailer. The girl experienced mosquito bites at night, and fly bites and wasp stings during the day. She ate fresh trout caught by her parents which was thoroughly cooked. All the potable water was from their camp trailer water reservoir and originated from the municipal water supply.

Blood was drawn for a complete metabolic profile and complete blood count and neurology and infectious disease were consulted on the patient. Neurology requested that the patient get imaging of the head and neck. Before any radiology or laboratory tests were performed, the infectious disease doctor performed a thorough examination of the girl's scalp and nape. Upon searching, a fully engorged tick was found attached at the base of the hairline at the nape (Fig. 5.4). The doctor carefully removed the tick with fine-tipped forceps, placed the tick in 70% ethanol, and submitted it to the microbiology laboratory for identification. Within 2 h of removing the tick, the girl began to regain function of her limbs and within 12 h all the symptoms had resolved. A clinical diagnosis of tick paralysis (TP) was made.

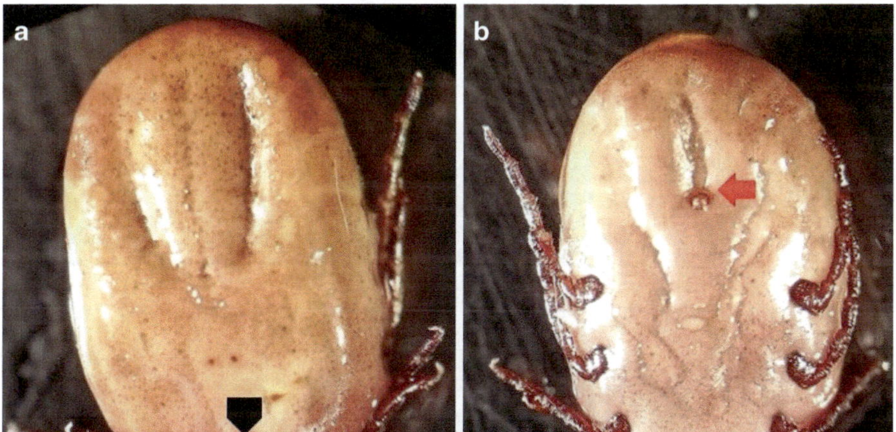

Fig. 5.4 *Dermacentor andersoni* dorsal view (**a**) showing dorsal shield (black arrowhead), eyes (black arrow) and (**b**) ventral view showing the anal groove (red arrow)

The tick was too engorged to appreciate whether festoons were present however an anal groove following the curvature of the body and eyes were visible (Figs. 5.4 and 5.5), which immediately ruled out *Ixodes* spp. The sex was confirmed by the fact that the tick was engorged, a feature only seen in a female tick (Fig. 5.4). The palps and hypostome (mouthparts) were short in relation to the basis capituli, and the dorsal shield was ornate, which supported an identification of *Dermacentor* (Fig. 5.5). Thorough examination of the spiracular plates provided species-level identification of *Dermacentor andersoni*, the Rocky Mountain wood tick (Fig. 5.6).

Discussion

Tick paralysis (TP) is an unusual medical condition that is associated with upwards of 40 species of ticks worldwide. Australia and North America have the highest recorded incidence of TP. In North America, *Dermacentor andersoni* is considered the most common cause of TP, with most cases being described in the pacific northwest (the United States and Canada), intermountain states (UT, ID, MT, WY, CO),

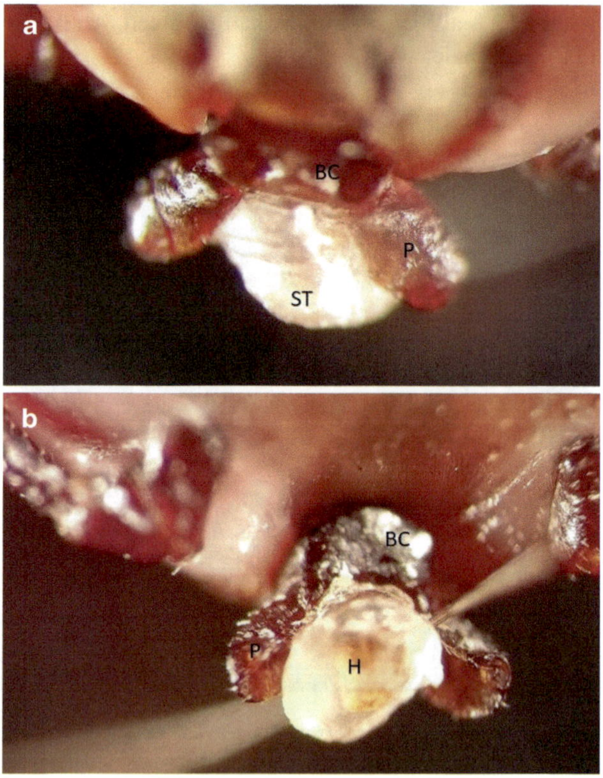

Fig. 5.5 Close-up view of the anterior end of *Dermacentor andersoni*. Dorsal view (**a**) shows the basis capituli (BC), palps (P), and scalp tissue from the patient (ST). Ventral view (**b**) shows the basis capituli (BC), palps (P), and yellow colored hypostome (H) embedded in the scalp tissue

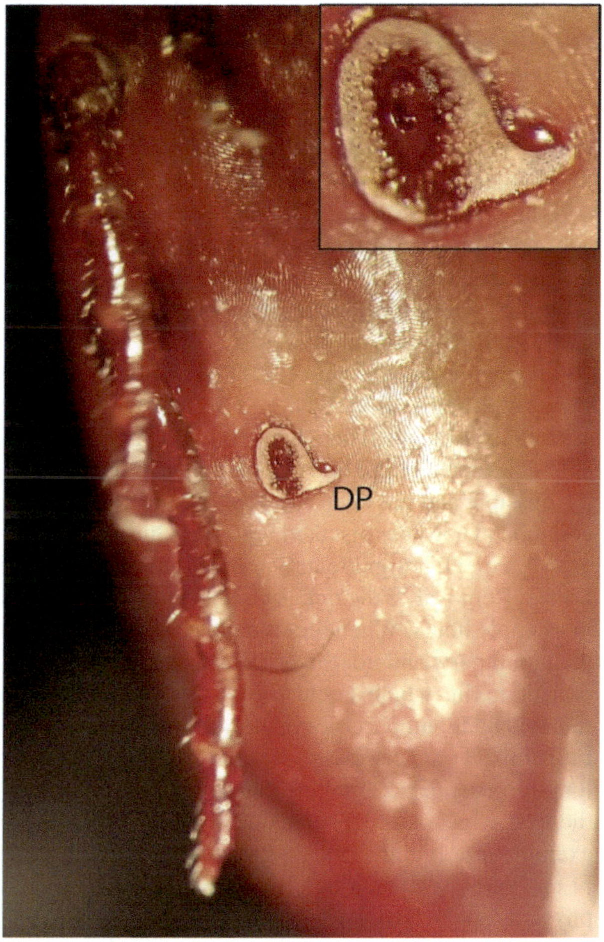

Fig. 5.6 Close-up view of the spiracular plate of *Dermacentor andersoni* on the left, lateral side of the body. Note the characteristic dorsal protrusion (DP) on the plate and few goblet cells present (small circles seen on magnified insert)

and Alberta, Canada. *Dermacentor variabilis* in the southeastern USA is also associated with TP, albeit less frequently than *D. andersoni*. Both *Amblyomma* and *Ixodes* can cause TP, but it is exceedingly rare.

Though the exact pathophysiology is not completely understood, TP is a result of an injected salivary neurotoxin that is produced by some females within certain species of hard and soft ticks. The conditions necessary for TP are restrictive in nature and dependent on the following conditions:

- The tick must be a female that has engorged and is capable of producing the salivary neurotoxin (not all females can produce the toxin).
- The female tick must have been feeding for 4–7 days total duration (mean of 5 days) in order for the toxin to be injected at a high enough concentration to cause clinical symptoms. *Ixodes* and *Amblyomma* tend to be more aggressive and voracious feeders, so they may be more apt to detach sooner than *Dermacentor*.

- The female must be feeding on the host at a site proximal to the central nervous system. Most cases describe recovering the tick behind the ears, at the hairline, on the scalp, or at the nape.

If all of the aforementioned conditions are met, clinical TP is able to progress. TP is most common in young children with long dark hair (though girls tend to have longer hair, thus are more classically associated with this condition). Having long dark hair makes it more difficult to detect a tick while it is attached and feeding, hence the duration of feeding and toxin injection is typically prolonged. The amount of toxin needed to cause paralysis is thought to be proportional to the size of the host, thus young children are generally more susceptible to symptoms of TP compared to fully grown adults.

The symptoms that precede TP are often vague and nondescript. Generally, the symptoms begin as fatigue and weakness progressing to ataxia and eventually ascending paralysis. There is no fever, rash, or altered mental status. If the condition is left untreated or the tick does not detach on its own, respiratory failure and death are possible and has been reported in upwards of 12% of untreated cases. The most common syndrome mistaken for TP is Guillain–Barré syndrome; however, the progression of TP is significantly more rapid and progressive than Guillain–Barré syndrome. Removal of the tick is curative and symptoms will typically begin to reverse in a couple hours. When removing the tick one should be careful to grasp the tick by the mouthparts and not the body, as squeezing the tick could result in a large injection of toxin, which could prolong or worsen symptoms. It is not necessary to remove the mouthparts completely, as the toxin is not maintained in the hypostome. Full clinical cure can be evident in fewer than 12 h, and it is highly unusual to have long-term sequelae. Imaging can sometimes be helpful in serendipitously detecting the tick as a small, irregular opacity that is then discovered upon further physical examination. Initial clinical assessments typically overlook TP as a potential cause of this paralysis, and a tick is typically detected by infectious disease or neurology physicians who have knowledge of this unique clinical presentation. Failure to consider TP in the initial examination can lead to unnecessary medical testing, prolonged hospital stays or unnecessary admissions, wasted healthcare expenses, and delayed removal of the tick.

Laboratory identification of the tick is not necessary for clinical management or diagnosis though determining if the tick is an engorged female within a genus/species associated with TP can be helpful to reaffirm the diagnosis. Many laboratories can provide basic arthropod identifications such as "hard tick" or "tick, not *Ixodes*" for the purpose of ruling in/out concern for Lyme disease specifically. More advanced genus and/or species-level identification is only provided by some larger specialized laboratories. Morphologic analysis is required to identify hard ticks to a genus or species level. Several features are evaluated to definitively identify the tick. In the case of *Dermacentor andersoni*, the presence of eyes and festoons, an ornate dorsal shield, short mouthparts, an anal groove following the curvature of the body, and a spiracular plate with pronounced dorsal prolongation and few goblet cells, allow for the definitive identification (Fig. 5.6).

Summary Points

- Tick paralysis is caused by more than 40 species of ticks worldwide, but its prevalence is highest in Australia and North America.
- *Dermacentor andersoni* is the most commonly associated tick with TP in North America and is caused by only a subset of engorged females.
- TP is characterized by rapidly ascending paralysis with preceding fatigue and ataxia.
- Removal of the tick is curative, with symptom improvement in a few hours. Full recovery can be achieved in as little as 2 h.

Selected References

Diaz JH. A comparative meta-analysis of tick paralysis in the United States and Australia. Clin Toxicol (Phila). 2015;53(9):874–83.

Halverson MR, Lagacé M, Smyth AE, Abu Sharar Z, Sargent MA. Imaging appearance of ticks in tick paralysis. Pediatr Radiol. 2021;51(3):493–4.

https://www.cdc.gov/dpdx/ticks/index.html

Simon LV, West B, McKinney WP. Tick paralysis. In: StatPearls [Internet]. Treasure Island, FL: StatPearls Publishing; 2022.

Case 3

James J. Dunn

A 9-year-old female who was otherwise healthy started to have headaches and vomiting episodes about a week after camping in south Texas. In the subsequent days, she began to experience other neurologic symptoms such as not being able to move both eyes in the same direction (gaze palsy), right-sided facial weakness, and a drooping right eye lid (ptosis). These symptoms persisted and she was finally admitted to the hospital when she became unsteady and uncoordinated while walking.

MRI of the brain showed abnormal signal in the brainstem and cerebellar peduncle and cerebellum extending across the midline (Fig. 5.7) suggestive of a demyelinating process although an infectious etiology could not be ruled out. A CSF obtained by lumbar puncture showed 321 WBC (normal range, 0–5/mm^3) with >80% neutrophils and three RBC (normal range, 0/mm^3). An extensive workup for an infectious etiology on testing of blood and CSF samples for bacterial, fungal, and viral pathogens yielded no conclusive answers. An additional CSF specimen was collected 2 weeks later and had 332 WBC with a predominance of neutrophils. This

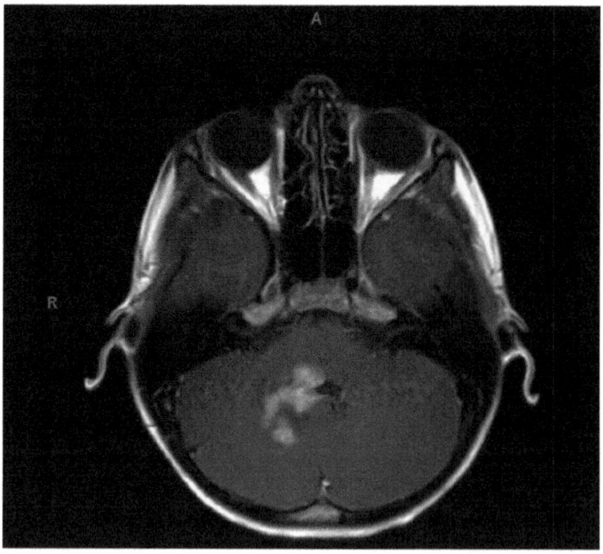

Fig. 5.7 Brain MRI showing abnormal signal in the cerebellar peduncle and cerebellum extending across the midline

Table 5.3 Differential diagnosis for chronic meningitis with CSF neutrophilic pleocytosis

Fungal pathogens	Bacterial pathogens	Noninfectious causes
Aspergillus	*Nocardia*	Systemic lupus erythematosus
Mucorales	*Actinomyces*	Chemical or drug-induced
Dematiaceous fungi	*Brucella*	
Candida	*M. tuberculosis*	
Blastomyces		
Histoplasma		
Coccidioides		

pattern of chronic meningitis with neutrophilic pleocytosis could include a number of infectious as well as noninfectious etiologies (Table 5.3).

After 6 weeks of persistent symptoms, the patient underwent a brain biopsy procedure in which fibrinous, pus-like material with black web-like consistency was obtained. Histopathologic examination showed pigmented, septated fungal hyphae associated with necrosis and acute inflammation (Fig. 5.8). A fungal culture of the material yielded a slow-growing mold with a velvety black surface with black pigment observed on the reverse side of the plate (Fig. 5.9a). Microscopic morphology of the culture isolate showed septate hyphae with long, sparsely branched, wavy chains of smooth, and oval conidia (Fig. 5.9b). Morphologically this organism resembled the dematiaceous fungi in the genus *Cladosporium*. However, the isolate did not produce the characteristic shield cells found with *Cladosporium* spp., and it was negative for gelatin hydrolysis (*Cladosporium* spp. are positive). Therefore, the fungal identification was reported as *Cladophialophora bantiana*.

The patient was treated for more than 1 year with three antifungal agents: amphotericin B, flucytosine, and voriconazole. She also underwent several neurosurgical

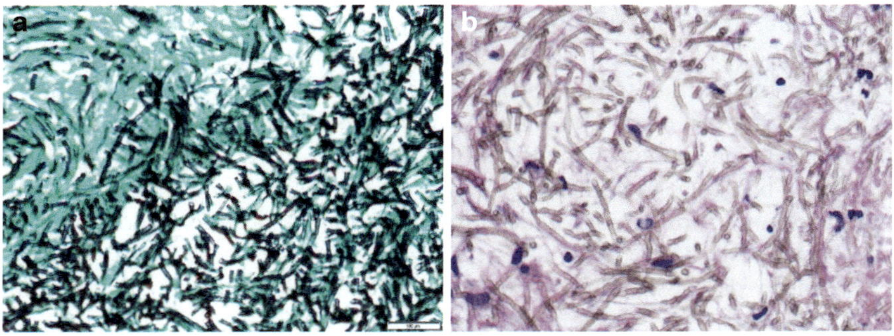

Fig. 5.8 Histopathology of the brain abscess showing septate fungal hyphae (**a**, methenamine silver stain, magnification 400×) that shows brown pigment indicating production of melanin (**b**, Fontana-Masson stain, magnification 400×)

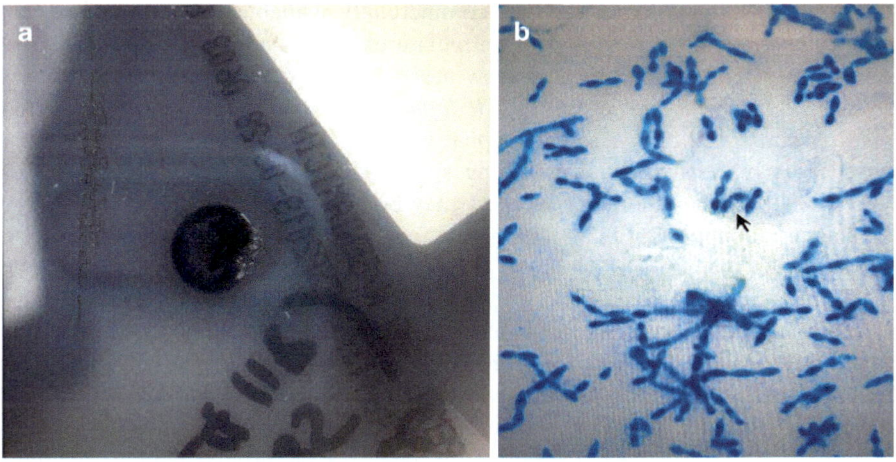

Fig. 5.9 Fungal culture of brain abscess. Colonies of *C. bantiana* (**a**) are brown or black and velvety. Microscopic morphology of the colony (**b**) shows septate hyphae with long, wavy chains of smooth, oval conidia

procedures to reduce the burden of organism present in the brain abscess. Her recovery from the infection has been complicated by continued cognitive and motor disabilities.

Discussion

C. bantiana is a highly neurotropic, dematiaceous fungus known to cause central nervous system (CNS) infection in both immunosuppressed and immunocompetent individuals. The organism can be found in soil and plant material in warm, humid environments and has a wide geographic distribution (Asia, North and South America, Europe, and Africa) with most reported cases from the Indian subcontinent. The portal of entry and route of infection in humans is not clear in most cases although inhalation followed by colonization and/or infection and subsequent hematogenous spread from an unrecognized pulmonary focus has been hypothesized. Higher rates of infection are seen in men, residents from rural areas, and those working as farmers, gardeners, and coal miners.

Laboratory diagnosis typically involves direct examination and culture of brain tissue, aspirated abscess contents and occasionally CSF. Histology findings generally show the presence of granulomatous inflammation with necrotic areas and microabscesses. The hyphae of *C. bantiana* are easily visualized with routine histologic stains such as H&E, GMS, and PAS. The Fontana-Mason stain will demonstrate melanin production with brown pigmented, septate, branched hyphae. *C. bantiana* will grow on most of the commercially available, general-purpose fungal media. Growth at 25 °C is slow with mature colonies apparent after 2 weeks of incubation. *C. bantiana* will also grow at 42 °C; a feature which distinguishes it from *Cladosporium* spp. and other *Cladophialophora* spp. The microscopic morphology consists of brown, septate hyphae with conidiophores that are similar to the vegetative hyphae from which arise long, sparsely branched, wavy chains of smooth, oval conidia. Since laboratory workers may be at risk of exposure to aerosolized organism, it is important that all manipulations of the fungus be done in a biologic safety cabinet.

There is no standardized therapy for *C. bantiana* CNS infection, but treatment typically involves a combination of surgical intervention and administration of multiple antifungals. There are no defined breakpoints for susceptibility testing but in vitro testing data suggest that voriconazole, posaconazole, flucytosine, and Amphotericin B have activity against this fungus. Prolonged treatment for months or years may be necessary. Despite these measures the prognosis of *C. bantiana* CNS infection remains poor, with case fatality rates over 60%, particularly in immunocompromised patients.

Summary Points

- *C. bantiana* is the most common dematiaceous fungi causing CNS infection.
- Diagnosis of *C. bantiana* infection is challenging since early brain imaging studies may appear to point to other etiologies.
- Despite aggressive surgical interventions and combination antifungal therapy, the prognosis for *C. bantiana* CNS infection is poor.
- *C. bantiana* grows slowly on routine fungal media and may be confused with *Cladosporium* spp. based on microscopic morphology.

Selected References

Chakrabarti A, Kaur H, Rudramurthy SM, Appannavar SB, Patel A, Mukherjee KK, Ghosh A, Ray U. Brain abscess due to Cladophialophora bantiana: a review of 124 cases. Med Mycol. 2016;54:111–9.

Kantarcioglu AS, Guarro J, De Hoog S, Apaydin H, Kiraz N. An updated comprehensive systematic review of cladophialophora bantiana and analysis of epidemiology, clinical characteristics, and outcome of cerebral cases. Med Mycol. 2017;55:579–604.

Velasco J, Revankar S. CNS infections caused by brown-black fungi. J Fungi (Basel). 2019;5:60.

Case 4

Marc Roger Couturier

A 54-year-old female from Montana presented to the emergency department at her local hospital with a minor scratch to her hand and wrist. The patient also carried a small cardboard box that contained a dead bat. On examination, the scratches looked superficial and were no longer bleeding. The patient recalled that she got into bed in the dark to avoid waking her husband who was already in bed. When she reached under the sheets, she felt a scratch and a small furry object. Thinking it was their kitten, she proceeded to touch it until she realized it was not the cat. After several small scratches and minor bites, she turned on the light, saw the bat, and proceeded to kill it with a shoe. She immediately boxed up the bat, washed her wounds with hot water and soap, and came to the hospital.

After consultation with the local health department's rabies virus (RabV) designate, the emergency department cleaned the wound with povidine-iodine and warm water. While the wound was being cared for, a rabies vaccine and human anti-rabies immunoglobulin were requested from the local health department as standard rabies post-exposure prophylaxis (PEP). The patient received both biologicals that evening and was held for observation until the following morning. The anti-rabies immunoglobulin was administered by injection in the hand, proximal to the wound. Remaining volume that could not be injected in the hand was administered to the bicep on the same arm. The vaccine was administered in the deltoid on the opposite arm. During the overnight period, it was discovered that the patient's husband had handled the dead bat to place it in a box while the wife was cleaning her wound. Based on this information, the husband also received PEP out of precaution.

The health department retrieved the bat from the emergency department and tested the brain by direct fluorescent antibody (DFA) stain for RabV antigen. The brain tissue was positive (Fig. 5.10), confirming that the bat was infected with RabV and that both patients were likely exposed. The couple both completed the full

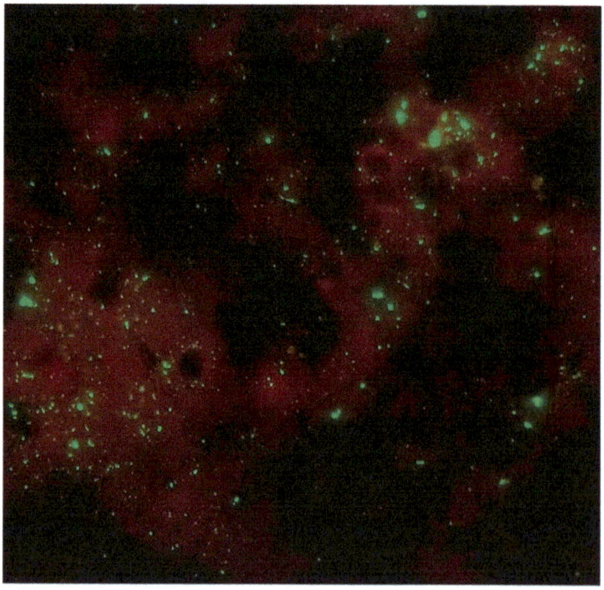

Fig. 5.10 Direct fluorescent antibody stain showing rabies antigen (green) in brain tissue (red) of an infected bat

course of vaccination and had an unremarkable clinical course without any symptoms.

Discussion

Rabies is caused by the prototypic viral species *Rabies lyssavirus* (RabV) and is a member of the genus *Lyssavirus* and family *Rhabdoviridae.* The virus has a unique bullet-shaped morphology under electron microscopy and contains a single-stranded negative-sense RNA genome. RabV is capable of infecting all mammals though there are only a few true reservoirs in nature for this virus. In the United States, raccoons, foxes, coyotes, and bats make up the major populations of reservoir host mammals. In Puerto Rico, the mongoose is the major reservoir host. Hawaii is the only US state without endemic RabV. The infection is spread almost exclusively through the bite of an infected mammal. Less common transmission events can involve direct mucous membrane exposure (without bite), laboratory aerosolization, and organ transplantation. Rabies is a gradually progressive neurological disease which typically begins with the virus entering the bloodstream of the host via a bite from an infected mammal. The virus replicates and spreads through the nerves with an incubation period that can range anywhere from 3 to 12 weeks. This period is highly dependent on several variables including: anatomic location of the bite (lower extremity bites progress slower than bites to the upper body), infecting dose, and any host immunity. During this incubation, the host is asymptomatic. The virus eventually reaches and terminates at the spinal cord and brain, where replication causes myeloencephalitis. At this point in the infection cycle, all infected hosts will eventually exhibit symptoms. The virus then spreads into the salivary glands, where it can be transmitted to a new host by subsequent bites and causes profound hydrophobia in the human host. Importantly, especially in domestic animals, the virus can begin to replicate and shed in saliva before clinical symptoms develop. The infected host typically dies within 7 days of symptom onset.

Infected humans rarely develop fulminant disease, thanks to preventative therapeutics available in most countries. If a human does progress to symptomatic disease, the initial symptoms will mimic flu-like symptoms with weakness, fever, and headaches lasting for multiple days. Itching or burning sensations at the site of the bite may occur. After several days of the flu-like symptoms, there is progression to cerebral dysfunction, anxiety, confusion, and agitation. As the disease further progresses, delirium, abnormal behavior, insomnia, and hallucinations will develop between day 2 and 10 of symptoms. In humans, once any clinical symptoms develop, the disease is nearly uniformly fatal. Fewer than 20 human cases of symptomatic rabies have reported survival. Of these cases, only a few did not receive pre-exposure (PrEP) or post-exposure prophylaxis (PEP). An important first step in managing suspected RabV exposure is thorough wound cleaning using a viracidal agent such as povidine-iodine. This cleaning should occur in any patient with suspected rabies exposure, regardless of vaccination history or documented titers.

The incidence of rabies deaths in domestic animals and humans in the United States has dropped significantly since the 1940s. This decline is largely in part to domestic animal vaccination and animal control policies, which has resulted in much less RabV in domestic settings, and fewer potential human interactions with infected animals. In nature, RabV is still maintained in host mammals; however, fatal human cases have been further limited by PEP using the human rabies vaccine and immunoglobulin therapy. The human rabies vaccine is derived from attenuated RabV and is recommended for both PrEP and PEP. The vaccine induces neutralizing antibodies and correlates of immunity by 7–10 days after injection. These antibodies are generally long-lasting (*i.e.*, detectable titers), but boosting is recommended in some specific at-risk populations to maintain correlates of immunity. In cases of suspected rabies exposure, vaccines are recommended as PEP in both vaccinated and unvaccinated individuals though the dosing frequency will be different.

Human anti-rabies immunoglobulin (HRIG) is another PEP that can be used to lessen the risk of developing rabies. These antibody injections are human IgG and allow for rapid neutralization of the virus during the first 0–7 days post exposure, preceding production of effective neutralizing antibodies from PEP vaccination. The HRIG should not be administered after day 7, as it can reduce the efficacy of the PEP vaccination. Likewise, in patients with previous vaccination and documented protective titers, HRIG should not be administered as it can also interfere with the natural immune response that would normally be effective 3–5 days after an infection.

Any patient suspected of exposure to rabies should always be given PEP immediately, and independent of any pursuit of laboratory testing. Laboratory tests on patients are generally not required if the biting animal can be tested. There are several tests available to diagnose rabies antemortem in humans. These tests are most useful when the patient is symptomatic, but they are limited to only a few reference or public health laboratories (including the Centers for Disease Control and Prevention, Atlanta GA). None of these methods are independently sensitive on their own, and thus a combinatorial approach is required if antemortem testing is pursued. Saliva can be tested by virus isolation in cell culture and/or nucleic acid amplification tests. Serum and spinal fluid specimens can be tested for antibodies to RabV (this is not the same testing as vaccine-response titers that are described below). Skin biopsy specimens can be examined for RabV by immunohistochemistry stains (antigen detection) in the cutaneous nerves at the base of hair follicles.

Rabies vaccine-response serology is primarily performed using the Rapid Fluorescent Foci Inhibition Test (RFFIT), which is a viral neutralization assay performed in cell culture and stained with a standard fluorescent antibody detection method. Rabies vaccine-response serology is never indicated as a diagnostic test. The correlate of immunity to rabies is defined as an antibody titer greater than or equal to 0.5 IU/mL though true protective immunity has not defined. Vaccination and laboratory testing for titers is only recommended for specific at-risk populations (Table 5.4).

When an animal bites a human, testing can be performed on the animal's brain (as part of a necropsy) to determine the presence of RabV. All wild animals involved in human bites should be euthanized and their brains examined for the presence of RabV by qualified rabies diagnostic laboratories. This testing requires brain tissue to be sampled from the animal and stained with specific fluorescent-labeled antibodies for microscopic examination (Fig. 5.10). Many state or territorial health labs can perform this testing. In cases of domestic animals biting humans, the animal need not be immediately euthanized unless there is strong suspicion of rabies based on the animal's behavior and symptoms. Unvaccinated domestic animals may require additional consideration. In cases of isolated bites or unusual behavior in a vaccinated domestic animal, the animal can be boosted with a rabies vaccine, isolated, and observed for up to 45 days. If the animal does not develop any signs or symptoms of rabies, the animal can be released from isolation. The patient should have already received HRIG and vaccination during this period as PEP. If the animal develops rabies symptoms or worsening behavior, the animal should be euthanized by a qualified professional and the head sent to a qualified laboratory for diagnostic testing.

Table 5.4 Rabies preexposure prophylaxis recommendations by risk category—2022 US recommendations (adopted from Rao et al. 2022)

Risk category	Population	Primary PrEP	Long-term follow-up
1. Elevated risk for recognized and unrecognized exposures including unusual or high-risk exposures	Rabies researchers and vaccine production workers and rabies diagnostic laboratory workers	Primary vaccine course	Titers measured every 6 months and boosters administered if titer is below 0.5 IU/mL
2. Elevated risk for recognized and unrecognized exposures	People who frequently handle bats, have contact with bats, enter high density bat environments, or perform necropsies (either bat biologists or rabies sample collections)	Primary vaccine course	Titers measured every 2 years and boosters administered if titer is below 0.5 IU/mL
3. Elevated risk for recognized exposures, sustained risk	Those who interact with animals that could be rabid, occupational/recreational activities that involve contact with animals (including—veterinarians, technicians, animal control officers, and their students or trainees), individuals who handle wildlife reservoir species (e.g., wildlife biologists, rehabilitators, and trappers), cavers/spelunkers, and travelers to areas with increased risk for rabies and decreased access to PEP	Primary vaccine course	One-time titer follow-up between years 1–3 post-vaccination and boosters administered if titer is below 0.5 IU/mL *OR* Boosters no sooner than day 21 and no later than year 3 after primary vaccine course
4. Elevated risk for recognized exposures, risk not sustained	Includes individuals in the same category as 3, but with an expected duration of exposure less than 3 years (e.g., intern, individual one-time project or activity)	Primary vaccine course	No titers necessary
5. Low-risk exposure	Represents the majority of the US population	No vaccination recommended	N/A

Summary Points

- Rabies is caused by the prototypical *Rabies lyssavirus* and is capable of infecting any mammal though natural reservoirs can vary based on geographic location.
- Rabies in humans, if left untreated, is nearly uniformly fatal.
- Testing for rabies is limited to highly specialized laboratories and is not considered a necessary standard of care for rabies infections.
- Post-exposure prophylaxis should be administered in any suspected exposure.

- Rabies vaccine and human anti-rabies immunoglobulin is the standard post-exposure prophylaxis.
- Necropsy and DFA staining of the brain for the detection of rabies is common in suspected infected animals.

Selected References

Campos-Outcalt D. Vaccine update: the latest recommendations from ACIP. J Fam Pract. 2022;71(2):80–4.

Centers for Disease Control and Prevention, Information for Diagnostic Laboratories available at https://www.cdc.gov/rabies/php/laboratories/diagnostic.html

Ma X, Bonaparte S, Toro M, Orciari LA, Gigante CM, Kirby JD, Chipman RB, Fehlner-Gardiner C, Cedillo VG, Aréchiga-Ceballos N, Rao AK, Petersen BW, Wallace RM. Rabies surveillance in the United States during 2020. J Am Vet Med Assoc. 2022;260:1157–65.

Rao AK, Briggs D, Moore SM, Whitehill F, Campos-Outcalt D, Morgan RL, Wallace RM, Romero JR, Bahta L, Frey SE, Blanton JD. Use of a modified pre-exposure prophylaxis vaccination schedule to prevent human rabies: recommendations of the advisory committee on immunization practices—United States, 2022. MMWR Morb Mortal Wkly Rep. 2022;71(18):619–27.

Case 5

Audrey N. Schuetz and Santiago Delgado Fernandez

A 17-year-old female presented to her primary care provider with complaints of a 1-week history of intermittent fever and chills, painful joints, and a rash. She also reported nausea, vomiting, and one episode of diarrhea. She works at a pet store and, notably, had been bitten by a pet rat at the store a few days prior to presentation. The patient judged the rat bite incident to be minor, and she reported barely noticing a wound on her hand the following day. On examination, a maculopapular rash was noted on the patient's palms and soles. No skin lesion was visible around the area of the rat bite. Both knees, elbows, and wrists were painful to the touch but were not erythematous or swollen. The patient was afebrile. Blood cultures were drawn, given the patient history of fevers and a rash. Laboratory analysis revealed mildly elevated white blood cells at 10.5×10^9/L (normal range, 3.4×10^9 to 9.6×10^9/L) with neutrophilia. After 48 h of incubation, the aerobic blood culture bottle flagged

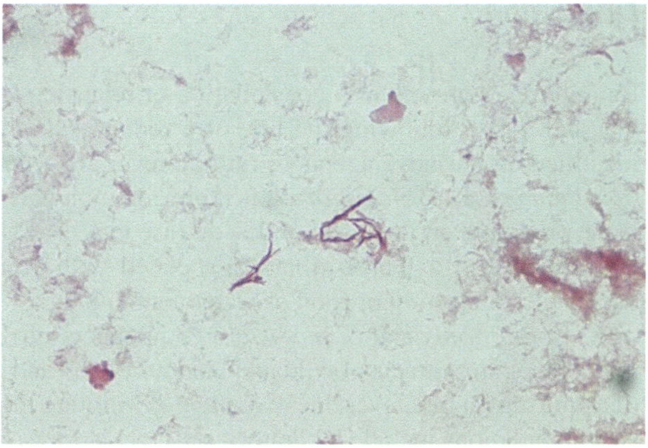

Fig. 5.11 Gram stain of blood culture bottle positive for filamentous Gram-negative bacilli of *Streptobacillus moniliformis* (1000× magnification)

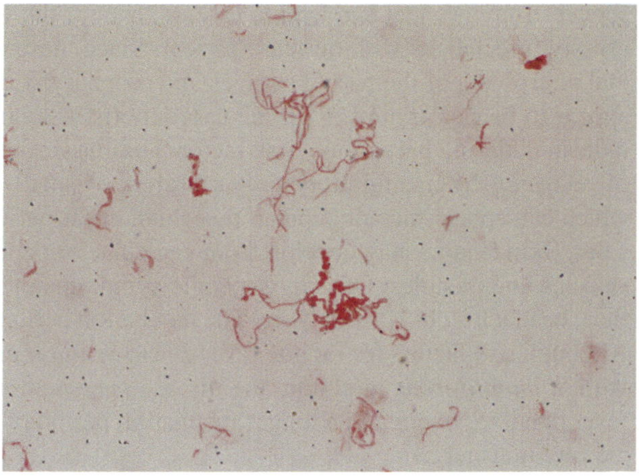

Fig. 5.12 Gram stain of colony from aerobic sheep blood agar plate showing filamentous Gram-negative bacilli of *Streptobacillus moniliformis* with bulbar swellings along the length of the bacteria (1000× magnification)

positive for growth. Gram stain of the broth showed filamentous Gram-negative organisms which were clumped together in tangles throughout the smear (Fig. 5.11). The patient was called to the hospital for further workup and treatment, given the Gram-negative bacteremia. Piperacillin-tazobactam was started empirically. After 18 h of incubation, small, gray, smooth colonies grew on the sheep blood agar plate incubated under aerobic conditions. Gram stain performed on colony growth showed fusiform Gram-negative rods with lateral bulbar swellings in tangled configurations (Fig. 5.12). Matrix-assisted laser desorption ionization-time-of-flight mass spectrometry (MALDI-TOF MS) identified the organism with high confidence as *Streptobacillus moniliformis*, a causative agent of rat bite fever. No antimicrobial susceptibility testing was performed on the isolate. The patient's symptoms improved, with no further fevers and disappearance of the rash and polyarthralgias. Antimicrobial therapy was switched to ceftriaxone intravenously for 14 days.

Discussion

Rat bite fever present as a constellation of relapsing fevers, migratory polyarthralgia, and a rash following the bite of a rodent. The two most common organisms identified as causative agents of rat bite fever are *Streptobacillus moniliformis* and *Spirillum minus*. *S. moniliformis* is more common in North America, while *S. minus* is the predominant pathogen of this disease in Asia. *S. moniliformis* can also lead to Haverhill fever, a febrile illness associated with vomiting and headache, which occurs after ingestion of food or water contaminated with rat feces.

S. moniliformis and *S. minus* are components of the normal microbiota of the upper respiratory tracts of wild and laboratory rats and other rodents. Persons at risk of acquiring rat bite fever due to *S. moniliformis* are those in close contact with rats; such populations include laboratory staff, pet store workers, pet rodent owners, and people living in unsanitary conditions with exposure to these animals. A bite or scratch from a rodent can lead to disease, even if the wound does not appear to be grossly infected, as with our case patient. Since these organisms comprise the normal microbiota of the nasal passages and oropharynx of the rat, the rodent may not appear to be sick or unwell. Some case reports of acquisition of rat bite fever have indicated kissing pet rats as a risk factor. The classic disease presentation is the triad of relapsing fever, migratory polyarthralgia of small and large joints, and a rash which can appear maculopapular, petechial, or purpuric, and can involve palms and soles. Skin lesions may develop from petechiae or papules into tender hemorrhagic vesicles and pustules. Other nonspecific symptoms include nausea, vomiting, diarrhea, headache, chills, and rigors. The incubation period is typically less than 7 days. Although risk factors for rat bite fever due to *S. minus* are similar to those associated with *S. moniliformis* infection, the disease presentation and course differ. Rat bite fever due to *S. minus* has a longer incubation period of 1–3 weeks, and polyarthralgias are unusual.

S. moniliformis is a Gram-negative, filamentous aerobic rod. On Gram stain, it can appear highly pleomorphic depending on the age and incubation conditions of the culture, ranging from straight to fusiform rods with characteristic bulbar swellings, or linear and tangled groups of rods. The Latin word moniliformis, from which this organism obtains its name, translates to "necklace-shaped," which alludes to the morphology of the bacterium on Gram stain (Fig. 5.12).

Classically, *S. moniliformis* has been described as extremely fastidious, requiring microaerophilic conditions and agar or broth enriched with 20% blood, serum, or ascitic fluid for growth. Many textbooks refer to inhibition of growth of this organism by sodium polyanethole sulfonate (SPS), an anticoagulant commonly used in commercial blood culture bottles. However, there have been many case reports of *S. moniliformis* recovered from modern blood culture media which contain SPS. Likewise, recovery of this organism is possible from standard culture media under routine aerobic incubation conditions, as occurred in this case. On agar plates, the colonies are gray, smooth, and glistening. In broth media, growth of this organism has been described as like a cotton ball or a "puff ball" in appearance, with growth most apparent near the bottom of the vial. Once isolated in culture, the organism can be identified by MALDI-TOF MS or 16S rRNA gene sequencing. Broad range bacterial 16S rRNA gene PCR and sequencing has also been used to detect this organism directly from a patient specimen if the organism is not recovered by culture.

The other agent of rat bite fever, *S. minus*, is also a Gram-negative rod but is uncultivatable. Diagnosis of rat bite fever due to *S. minus* is often made empirically in patients with sepsis and a history of rat exposure.

Treatment recommendations include penicillin or ceftriaxone as the drugs of choice, but tetracyclines and other cephalosporins are also active against the bacterium. The mortality rate of untreated cases reaches 10%. Because of the nonspecific nature of the clinical presentation, the differential diagnosis is broad, including but not limited to meningococcemia, bacterial endocarditis, and toxic shock syndrome. Prompt diagnosis of this entity requires an astute clinician to question the patient regarding rodent exposure.

Summary Points

- A bite from a rodent is a risk factor for rat bite fever due to *Streptobacillus moniliformis*.
- Population groups at risk for rat bite fever include laboratory workers, pet store workers, and other persons in close contact with rats and other rodents.
- *S. moniliformis* is classically described as fastidious in growth, but recent reports have shown recovery in standard culture conditions.
- Because of the nonspecific nature of the clinical presentation, high clinical suspicion is helpful in achieving organism identification and timely treatment of the disease.

Selected References

Elliott SP. Rat bite fever and *Streptobacillus moniliformis*. Clin Microbiol Rev. 2007;20(1):13–22. https://doi.org/10.1128/CMR.00016-06. PMID: 17223620; PMCID: PMC1797630.

Fukushima K, Yanagisawa N, Imaoka K, Kimura M, Imamura A. Rat-bite fever due to *Streptobacillus notomytis* isolated from a human specimen. J Infect Chemother. 2018;24(4):302–4. https://doi.org/10.1016/j.jiac.2017.10.018. Epub 2017 Nov 27. PMID: 29191371.

Ojukwu IC, Christy C. Rat-bite fever in children: case report and review. Scand J Infect Dis. 2002;34(6):474–7. https://doi.org/10.1080/003655402320170345. PMID: 12160180.

Rodino KG, Miller NE, Pethan KD, DeSimone DC, Schuetz AN. The brief case: rat bite fever from a kiss. J Clin Microbiol. 2020;58(1):e00677-19. PMID: 31871056 PMCID: PMC6935932.

Index

J. J. Dunn et al., *Challenging Cases in Diagnostic Clinical Microbiology*,
https://doi.org/10.1007/978-3-031-65092-5

C

N

O

P

GPSR Compliance

The European Union's (EU) General Product Safety Regulation (GPSR) is a set of rules that requires consumer products to be safe and our obligations to ensure this.

If you have any concerns about our products, you can contact us on ProductSafety@springernature.com

In case Publisher is established outside the EU, the EU authorized representative is:

Springer Nature Customer Service Center GmbH
Europaplatz 3
69115 Heidelberg, Germany

Batch number: 10755227

Printed by Printforce, the Netherlands